BAMBOO

BAMBOO

by ROBERT AUSTIN
and KOICHIRO UEDA
photographs by DANA LEVY

WEATHERHILL
New York & Tokyo

First edition, 1970
Sixth printing, 1978

Published by John Weatherhill, Inc., of New York and Tokyo, with editorial offices at 7-6-13 Roppongi, Minato-ku, Tokyo 106, Japan. Copyright in Japan, 1970, by Robert Austin, Dana Levy, and Koichiro Ueda; all rights reserved. Printed in Japan.

LCC Card No.70-96051
ISBN 0-8348-0048-9

Contents

Bamboo: Its Lore and Versatility   9

Bamboo: Its Beauty and Uses   23
    In Nature   24
    In Garden and House   44
    In Everyday Use   88
    In Crafts and Art   128

Bamboo: Its Growth and Cultivation   193
    Growth   193
    Planting and Cultivation   195
    Flowering   199
    Industrial Cultivation   201
    Special Techniques   205
    Tables of Species   209

Photo Notes   211
Bibliography   213
Acknowledgments   215

Bamboo: Its Lore and Versatility

Bamboo is one of the most extraordinary plants that exist. It flowers perhaps once in a hundred years, and then it dies. It grows faster than anything in the world. In fact, it is sometimes possible to *see* it growing, just as one can see the hands of a large clock moving: there are recorded instances of bamboo's growing four feet in a single day. In a grove in spring the vitality of the surrounding green pillars is almost palpable. While the stem is growing above ground, the root stops: when the stem has finished, then comes the turn of the other. Bamboo also possesses the characteristic of making its complete growth in about two months only. Thereafter it remains the same size as long as it lives.

But bamboo is interesting for much more than this: it is the most universally useful plant known to man. For over half the human race, life would be completely different without it. The East and all its peoples can hardly be discussed without bamboo's being taken into account. Accepted as a mere fact of life or prized for aesthetic reasons, it touches daily existence at a thousand points which vary as widely as its employment in literary metaphor and its use in the walls of houses. It serves the most mundane purposes, and the most refined: dwellings are constructed from bamboo; it is widely used for eating and drinking utensils and for countless other household implements. Ubiquitous, it provides food, raw materials, shelter, even medicine for the greater part of the world's population. The interlocked roots of a bamboo grove restrain the river in flood and during earthquakes support the insubstantial dwellings of country villages.

As a traveler of the Victorian era, Colonel Barrington de Fonblanque, observed, a little pessimistically: "What would a poor Chinaman do without the bamboo? Independently of its use as food, it provides him with the thatch that covers his house, the mat on which he sleeps, the cup from which he drinks, and the chopsticks with which he eats. He irrigates his field by means of a bamboo pipe; his harvest is gathered in with a bamboo rake; his grain is sifted through a bamboo sieve, and carried away in a bamboo basket. The mast of his junk is of bamboo; so is the pole of his cart. He is flogged with a bamboo cane, tortured with bamboo stakes, and finally strangled with a bamboo rope."

They call it *také* in Japan, in China *chu*. The Western nations use the word "bamboo," with slight variation from country to country, and one of the earliest references to it, in Hakluyt's *Navigations*, notes that all the houses in Indonesia "are made of canes, which are called Bamboes, and bee covered with straw." Two centuries later, the first edition of the *Encyclopaedia Britannica* still retained this spelling. The origin of the word is unclear: the accepted explanation proposes agreeably enough that it derives from an onomatopoeic Malay word imitating the explosive noise bamboo makes when burning. Marco Polo also observed how travelers would tie several green poles together and suspend them near the campfire: exploding at intervals through the night, they would frighten away marauding beasts.

The Chinese and Japanese write bamboo as 竹, representing two plants side by side with a branch and leaves dependent. Now stylized, the symbol was originally a pictograph which has been simplified by the passing of the centuries and the flexible demands of the writing brush. Indeed, the symbol for the writing brush itself has naturally for over two millennia included the character for bamboo—as has also the word for gambling, on account of bamboo's associations with cards, dice, and mah-jongg. From time immemorial in China, books were inscribed by hand on strips of bamboo linked with silk; there are still in existence bamboo strips two thousand years old which detail military provisions.

In the Japanese city today, bamboo can be found outside homes, where a length will be carved or painted to show the name of the owner, or inside elegant teahouses, where two pieces incised with a complementary poem hang on either side of the door to bring good fortune. A restaurant may exhibit a few discreetly lit plants of greater or lesser size to evoke the tranquillity of the country. A shoot or two tap on the window of even the most modest hotel. And for New Year decorations, outside many houses there stand small viridian bundles of pine branches, topped with three lengths of bamboo freshly cut from the greenest stems, mollifying the pavement's severity by their rural presence.

In its more sophisticated and exotic applications also, bamboo is inextricably woven into the fabric of Eastern culture. Bamboo flutes have existed almost as long as the human race. The tea ceremony and the art of flower arranging would hardly be possible without it, nor would the ritual and delicacy which come together to produce the fan and umbrella in all their many varieties. The paper of a fan may well be made from bamboo too, paper which can also become the sliding wall of a room or be lettered with an ephemeral and allusive poem. The people of Vietnam epitomize its closeness in their proverb "The bamboo is my brother."

Bamboo has been directed in the past as now to unexpected tasks that link the East and West. Gramophone needles were made for connoisseurs from selected pieces; Edison, after an infinity of vain experiment, found bamboo ideal as the filament for his first electric lamps; in Manila there is a church organ a century and a half old with pipes of bamboo—and a similar organ, with bamboo pipes, was recently built and installed in one of central Tokyo's modern buildings. The early Dutch traders in the Far East rapidly adopted a practical device they found the local inhabitants using in their beds, an open bamboo framework that allowed air to circulate around the body on hot summer nights; originally called a "bamboo princess," this soon became known as a "Dutch wife." Bamboo can also be used for making fire: a piece of porcelain struck against rough-surfaced bamboo will produce sparks, or a hollowed-out length of bamboo makes an ideal "fire piston"—a tube with a tightly fitting plunger which heats air by compressing it quickly, thus igniting the tinder.

The design of the ship itself in the East derives from bamboo. In the West, the first boats were dug-out canoes and all Western shipbuilding has followed in this pattern. Oriental ships, however, descend from the ancient bamboo sailing raft—which is still in use—and are built in a box-like construction. The many battens in their sails are still made of bamboo. The accompanying concepts of the watertight compartment and the stern-post rudder were adopted many years later by the West.

In his history of the Chinese junk, G. R. G. Worcester catalogues the uses to which the ingenious junkmen put bamboo: "rope, tholepins, masts, sails, net floats, basket fish-traps, awnings, food baskets, beds, blinds, bottles, bridges, brooms, foot rules, food, lanterns, umbrellas, fans, brushes, buckets, chairs, chopsticks, combs, cooking gear, cups, drogues, dust pans, paper, pens, nails, pillows, tobacco pipes, boat hoods, anchors, fishing nets, fishing rods, flagpoles, hats, ladders, ladles, lamps, musical instruments, mats, tubs, caulking material, scoops, shoes, stools, tables, tallies, tokens, torches, rat traps, flea traps, joss sticks, and back scratchers."

The building of bridges in the Orient is largely based on the use of the bamboo cable. Over the Min River in China there was, and still is in all probability, a bridge 250 yards long made solely from bamboo without nail or iron. It is 9 feet wide and supported by 20 cables, half under the treads and the rest forming the sides. These cables, 21 inches in circumference, are made solely of bamboo split and twisted together. Similar cables, though much smaller, are still used for towing junks through the terrifying rapids of the Yangtze River—where up to three hundred men may pull the tow ropes. Experts have estimated that the working stress of these lines exceeds ten thousand pounds per square inch. Bamboo is preferred to hemp because it is far stronger on a straight pull, is lighter, and better resists the severe friction over the rocks. The lines are available in lengths almost a quarter of a mile long, one of which cost the equivalent of $15 in 1940.

To Western eyes, the charm of most bamboo articles is immediate, deriving from two sources—their modesty and their suitability. Most bamboo objects hardly seem manufactured; they appear more as products of nature. Shaped and refined over the years for unchanging everyday needs, their form and feel are satisfyingly appropriate—indeed inevitable. They give the impression of having generated themselves, having evolved slowly for their tasks, without acquiring complexity from man.

One cannot, for example, easily put out of mind the purity of line and the consummate simplicity of even a water ladle. Here are two pieces only of bamboo, almost in their natural state, held together by nothing more than their swelling power in the water which becomes the ladle's element for the greater part of its life. But its beauty—a quality that age has conferred on an object which initially was merely correct for its purpose—brings a sense of excitement to the eye. This same quality is reflected in a hundred everyday things, such as chopsticks, or a tea whisk, the first of small consideration, the second highly appreciated, but both created out of a long tradition with nothing more than a knife as tool, and from material of little cost. Or one may note the baskets to be seen by thousands in markets—flimsy, ephemeral confections produced to hold a few oranges, some radishes, or perhaps to bring hens for sale, and made usually by an old couple, their only tool an ancient and heavy splitting-knife. These airy nothings seem to grow with a swiftness of their own, almost independent of the fingers that fashion them with an exactness made automatic by repetition. Like a bird's nest, like the honeycomb, merely to view them gives pleasure from their perfection of form—a perfection produced unawares in a humble object through an evolution more or less instinctive. Consider also the bamboo baskets made for flower arrangements by the masters of this art. Their refinements are often not to be perceived by the Occidental eye and they may appear as merely baskets; but, for a small group with the knowledge to understand and the means to purchase, an exquisite degree of sophistication reveals itself. A contrast to this connoisseur's approach may be found in the twenty-five feet of split, tapering bamboo, swiftly twisted into a springy circle and hammered over a fishtub, which is thereafter able to withstand for years the punishment it gets on trawler and quayside.

Regarded as a material to work, bamboo shows itself "grateful"—to use the artisan's term. It is flexible yet tough, light but very strong. It can be split with ease, in one direction only, never in the other; it may be pliant or rigid as the occasion demands; it can be compressed enough to keep its place in holes; after heating, it can be bent to take and retain a new shape. It is straight and possessed of great tensile strength.

Nor should the usefulness of the bamboo sheath be overlooked. Wrapped tightly round each joint of the stem until it falls or is cut off, this large leaf-like covering sheet has many applications where a light strong covering sheet is called for—from hats and roofs to fans and food-wrapping.

The leaves of bamboo seem never to be still, and after rain they become a million flashing mirrors. In the hottest weather their rustle lends an illusion of coolness—and of endless whispering. This is often heard as melancholy by Eastern poets: a thousand years ago Ou-yang Hsiu wrote how the bamboos tap out the music of autumn, "Myriad leaves give a thousand sounds—all are lamentation." In the present century, William Plomer wrote:

> I love, bamboo, your fidgets
> And sudden sighs, bamboo;
> Awake alone I listen

> To secret susurration
> Like paper scraping stone,
> Stroking the inner surface
> Of this old heart, bamboo.

From the botanical point of view, scholars still argue about the classification of bamboo. It is usually catalogued in the grass family, but has also been labelled as a kind of rice because of its flowering properties. However, most authorities now agree that bamboo is unique. Its family, known as *Bambusaceae*, is distinguished by: the special structure of its stem or culm; the fact that it reaches its full height in a short period; its rapid rate of growth; its singular flowering habits.

The characteristics of bamboo demand the use of two special words. The term culm is used in place of trunk, which fits trees well enough but is inappropriate in this case since the culm grows so rapidly, has joints, and is hollow. Similarly, the generic "root" suits bamboo uneasily. Bamboo has indeed a very large growth below ground and also small roots proper, but what distinguishes it—or at least half the species—is the rhizome. This is a long and fast-growing underground shoot from which germinate the new sprouts. It is the speed with which the rhizome travels underground that accounts for the spacing of the culms, each of which possesses its own system of small roots.

But one must be more precise. There are two main types of growth. Bamboo is found either in clumps of culms or as single, free-standing culms. The first pattern of growth, known as sympodial, is widespread in the tropics; the second, called monopodial, is usually found in cooler climates, like that of Japan. The two patterns are determined by the growing habits of the rhizome. In the monopodial type the rhizome travels very far; depending on the species, it will often grow over a hundred yards underground. In a grove the amount of rhizome intersection is formidable, and this is the reason why such a subterranean network holds the ground so firmly during earthquakes. Each joint or node of this far-spreading root bears a single bud yearly and some of these germinate to grow through the surface and become new bamboo. These single sprouts—in Japanese, *takenoko* or "bamboo children"—already possess in emerging the final diameter of the adult culm. The culms are erect and are clear of branches for a considerable height. The sympodial type, also known as clump bamboo, is less diffuse in its growing habits. Here too the rhizome puts forth an underground shoot, but this is connected directly to the parent and the link is short. Thus, the mother bamboo keeps her progeny close around her in a tight group. Sprouts of clump bamboo appear a month or more later than the monopodial type—in July or August, depending on the region. The accompanying figures show the difference between the two types of growth.

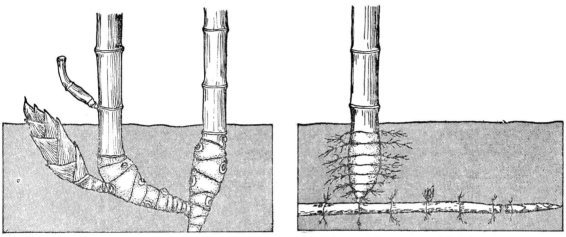

BAMBOO GROWTH PATTERNS: SYMPODIAL (left) AND MONOPODIAL (right)

There is also a kind of bamboo, called sasa, which is more nearly related to grass and grows only to a modest height, sometimes a matter of a few inches. It is definitely a part of the bamboo family and covers, for example, millions of acres of Hokkaido in the north of Japan. Sasa retains all its life the sheath which grows round the node or joint; most other types of bamboo drop it in growth. Its green and white leaves are used in Japan to decorate dishes of raw fish.

The theory is that bamboo originally existed only in the monopodial form, which continues growing throughout the year, periods of rhizome growth alternating with those in which new sprouts and culms develop. In the tropics, however, where the rainy season is followed by a long drought, lack of water makes such a growth pattern impossible. Sympodial bamboo probably developed in response to such unfavorable conditions. Here, rhizome and culm have become one, so that the bamboo can accomplish all its growth during the rainy season and then rest for the remainder of the year. Clump formation in the tropics is also favored by the fact that bamboo generally prefers some shade—which is naturally available in a group of culms.

Further, culms grouped closely together offer mutual protection against damage.

Bamboo flourishes in climates which are warm all the year. There are no native bamboos in Europe and North America, and very few in Australia. It is in Asia that bamboo truly comes into its own, particularly in Japan, China, and India. The northern limit of its survival is a little above the fortieth degree of latitude, which runs through northern Japan, Vladivostok, and Peking. Here temperatures of zero degrees Fahrenheit can occur, which spell extinction to all but a few very hardy types of bamboo.

In all types of bamboo, the most striking characteristic is immense vitality. Bamboo, it seems, can overcome almost any kind of hardship. With its far-ranging network of growth beneath the ground, all shoots are linked together and nourish each other and propagate apparently without end—without end, that is, until bamboo flowers. This coming into flower of bamboo represents the swan song of its existence. It happens infrequently, depending on the species, at such long intervals as every 60 or even 120 years; and then most plants perish. The flowering may happen only once in several human generations, and the occurrence (as with the brighter comets) is habitually taken by people as presaging some great disaster. As a portent, this event is made more impressive in that the flowering of one type occurs almost simultaneously all over the country—and even in those cuttings transplanted into other lands. This is regarded as satisfying proof that the genus sprang originally from the same stock at some period now obscured by the mists of time. In fact the flowering is usually spread over a year or two, but at the date of this writing, the madake, a long-jointed bamboo, has bloomed in Japan, an event which last occurred before the Civil War in America and before the Black Ships of Commodore Perry opened Japan to the West. This particular bamboo accounts for nearly three-quarters of that growing in Japan, and its dying will be a serious blow to the bamboo confraternity, as it may be another ten years before the groves come back to their former state.

In bamboo cultivation the most important matter is the choice of the "mother bamboo." Since the individual plant does all its growing in a single season, the function thereafter of the culm is not to gather nourishment for itself but for the new sprouts, to which the parent also transmits its characteristics. For this reason, the name mother bamboo is highly appropriate. If culms of a large size are wanted, then a young and substantial mother should be selected. If on the other hand finer stems are necessary, then a slimmer and younger mother, three or four years old, must be chosen. All that is needed is to fertilize these maternal culms every spring and summer. This has no effect on the mother, but it is of great and immediate benefit to the rhizomes which feed the young shoots that grow up from them. Groves in mountainous areas are better than those of the plains in yielding good bamboo. The best bamboo for industrial use is that with long joints having relatively thin walls and some flexibility.

The habits of bamboo are often singled out for Oriental precept. The straightness and speed of its growth is noted in an allusive Japanese haiku:

> A young bamboo —
> how tall it has grown,
> without the slightest help in the world.

That the plants live in a large family, close together yet proud and independent, is often singled out for comment. Much is also made of the unselfishness of the parent bamboo. When the mother culm has her family standing round her, she no longer accepts sustenance herself but passes it all on to strengthen the young shoots. But the most well-known exemplar afforded by the bamboo is the analogy to a wise and patient man who knows how to bend before the winds of life and is never broken. One of the favorite subjects of the screen paintings in medieval Japanese castles was a tiger in a bamboo jungle: there the swift directness of the golden tiger found its complement in the straightness of the immobile green stems.

The life cycle of the bamboo is defined and regulated. From the moment that the sprout appears through the soil, it undergoes no increase in diameter, and in it are visible all the joints that will constitute the grown culm. The fact that the bamboo plant proceeds from birth to maturity in less than a year means that the schedule for cutting can be quite regular and much more frequent than in the case of trees grown for timber. It is best that groves be cut every year; if the bamboo shoots are wanted for food, then they must at times be tended almost daily. Harvesting yearly enables the culms to be selected and cut at the right age. It is not always possible to find enough labor for an annual harvest, however, and a compromise often has to be made. In general, age brings little benefit and there is no advantage in leaving bamboo uncut after five or six years. Extremely young bamboo tends to be too soft for manufacturing purposes, while the older plants become too hard and are also prone to damage by insects. The most useful culms are three to five years old with thick walls and the smaller and more flexible stems with thin walls.

Bamboo is much esteemed as food in the East. Usually cooked by frying lightly, it is crisp and pleasant, in texture a little like an apple and in taste resembling artichoke heart, and its food value corresponds roughly to that of the onion. Only the young shoots are eaten, and they must be dug up without delay since leaving them in the ground for even a short while will make them large and coarse. (A little time may be gained by banking up the shoot with earth or a basket to protect it from the light.) Soon after the sprout appears through the soil, it should be gathered and stripped of its outer covering. Shoots are normally boiled to remove any bitterness before being sold. Apart from canning for export, dried bamboo shoots are also popular in China. They are processed by boiling in salt water, drying for four hours in a closed chamber, then pounding flat or slicing. The price varies according to their tenderness; ten pounds of fresh shoots produce a single pound only of dried shoots.

In any garden, Oriental or Occidental, the bamboo is both useful and pleasant. It makes a handsome hedge of considerable strength, and as an evergreen plant of attractive appearance, with the rustle made by its foliage in the slightest breath of wind, the bamboo pleases in many ways. Sasa, the small bamboo grass, is very decorative in a garden: it sets off rocks, paths, and fountains well and may also be trimmed freely into whatever shapes are appropriate. Tables 9 and 10, beginning on page 209, give details concerning species which can be grown easily in different weather zones; they are referred to below by their Japanese names.

The low kumazasa has large white-rimmed leaves, which evoke mountain scenery and accompany running water; the species with dotted leaves bring a note of brightness to the garden. Young shoots of yadake have single branches which grow very straight; when wrapped in their white sheaths they look extremely elegant. It is from this bamboo that arrows are made, *ya* meaning arrow in Japanese.

Shihochiku grows squarish in form with short pointed spikes at its joints, and is planted near gates to keep out evil powers. The black kurochiku is a slender and hardy plant which can often survive temperatures below freezing point. It is the first bamboo recorded as being imported into Europe, the date being given as 1827. Moso may be kept to a medium height in the garden; its leaves, being small and elegant, are cultivated for ornament. If its branches are trimmed at the top, new leaves grow thickly from them the following spring and by creating a visual barrier lend space to the garden.

For a small garden it is better to select the clump or sympodial type of bamboo where the young plants grow closely around the parent, although it does not in general like cold weather and not all species can be transplanted. One that can travel is suhochiku, a plant of golden culms with bright green vertical stripes and young shoots that are particularly beautiful.

Bamboo in a garden must be kept under control, particularly the monopodial type. The rhizome must not be allowed to spread unchecked; the old culms and unwanted sprouts should be cut every year. This is not only to prolong the life of the remaining culms but also to make room for new shoots.

There are over a thousand kinds of bamboo. The botanists put forward about 1,250 species, classified into approximately fifty genera. In Japan half of this number at the most can be found, but the common types number less than fifteen. In fact ninety percent of the bamboo in Japan comes from two kinds only—madake, which forms by far the greater part, and moso. The most usual botanical names for these varieties are *Phyllostachys bambusoides* and *Phyllostachys pubescens*. Madake is the most widespread bamboo in Japan and the one most generally used in industry; moso is best for eating and is known as the "noble bamboo." Both can grow in warm regions (though not in the tropics) to heights of over sixty feet, with diameters of seven inches.

A distinction is often made between male and female bamboo, but the definition differs widely according to who makes it. The Japanese craftsmen who live and work with bamboo label as male that which is thick and tough in quality and has a short joint. The "female" material is thinner, more flexible, and has a longer length between the nodes. Another definition is made from the number of branches at the first joint of the culm. If there is one branch, then the bamboo is considered male; if two, female. Others say that male bamboo has deep grooves on its surface and also loses its sheaths as soon as branches grow. All this apart, there is no sexual difference between the species.

Since bamboo is such an integral part of daily life, its folklore is almost that of the East itself, and the legends and stories built around it are innumerable. One of the oldest in Japan tells of the Shining Princess Kaguyahime. This is about a poor and childless farmer who one day came across a magnificently tall bamboo and, cutting it down, found a small girl standing inside the culm. He took her back to his house and cherished her; she grew up to be a beautiful and accomplished lady. The fame of her beauty having spread far and wide, five noblemen competed to win her hand in marriage. She remained unmoved by their proposals, setting each suitor such impossible tasks that all were at last obliged to give up. Finally, in the best tradition, the emperor himself fell in love with Kaguyahime. This devotion from such a high place brought her embarrassment and sadness. She declared that since she never wished to marry, angels would come down from the moon when it was at its fullest and carry her there—her original home. On hearing this, the emperor despatched two thousand soldiers to stop her flight. But it was in vain, and she escaped into the heavens. She left a letter, which the emperor read and ordered to be burned with solemnity at the top of the highest mountain in the land. This was done, and the smoke magically persisted; hence the mountain came to be known as Fuji, the immortal one.

The incident of the seven wise men in the bamboo grove forms an important part of Chinese cultural history. Tired of the pomp of court life, these philosophers decided in the third century to retire from the world and immure themselves in a bamboo thicket to discuss culture and philosophy without distraction for the rest of their lives. They did this, and their practical example had great influence on succeeding generations.

To history also belongs one of the best-known incidents of bamboo bringing richness to the Western world. This was the occasion when the eggs of the silkworm were first smuggled in a bamboo staff from China to Constantinople in A.D. 552. The monks who carried them—at the risk of a most disagreeable death—thus initiated the decline of the great Silk Road that had spanned Asia for centuries.

Eight hundred years ago the Chinese poet Pou Sou-tung wrote: "A meal should have meat, but a house must have bamboo. Without meat we become thin; without bamboo, we lose serenity and culture itself." The seventeenth-century Japanese priest Gensai, living at Kyoto, wrote:

> Bamboo leaves hang in front of my house;
> At the back, they divide it from the world;
> They cover it above and give shelter.
> I, the bamboo lover, find home within their shade.

On his death he asked that bamboo should be planted instead of a tombstone. To this day, three bamboos are tended at the place where he is buried. In another garden of Kyoto (this old capital holds bamboo dearer than any other part of Japan) there grow seven small black bamboos on a tiny island. The garden was made six hundred years ago and its designer, Muso-kokushi, planted them that his seven favorite pupils should be remembered; the bamboo there today is descended from that originally planted by Muso-kokushi.

The tradition of cutting the umbilical cord of a newly born baby with a bamboo knife is still maintained in parts of the Chinese and Japanese countrysides. It is believed that the knife for a girl should be made from male bamboo while female bamboo should be used for a boy. According to legend, when Princess Konosakuya gave birth to a son, the knife that was used was thrust upside down into the ground; taking root, a strange form of bamboo grew up. It was called topsy-turvy bamboo, still found in Kagoshima Prefecture in Japan.

In China, bamboo is one of the four noble plants—the others being the orchid, the plum tree, and the chrysanthemum—and again and again through the centuries it forms the subject of pictures which use its straightness to illustrate a moral. It is also—with the plum and this time the pine tree—one of the Three Friends, bamboo representing Buddha, and the others Confucius and Lao Tzu. These three plants as a group also symbolize happiness and good fortune.

In Japanese crests, bamboo and sasa occupy a considerable place. Embroidered in miniature on formal kimono, the crest shows the family of the wearer. There are several thousand different crests, many being based on natural objects, prominent among which is the bamboo; the word for bamboo also forms a part of many Japanese family names.

In Oriental medicine also there are many remedies based on bamboo. Perhaps most of them are only superstitions, but some in fact seem to work and need study rather than ridicule from the medical profession. A case in point is Tabaschir, a preparation of widespread fame in the Middle Ages. This near-mythical potion was supposed to be an antidote to any poison and was derived from bamboo in a jealously guarded manner. Such credulity was thoroughly discounted—until it was recently found that inside the culms of certain bamboos there exists natural silica in a microscopically fine powder. And nowadays an artificial form is used internally to neutralize toxic agents by absorption. Other traditional treatments may yet turn out to be efficacious. One nostrum (recommended for such varied ailments as asthma, spitting blood, or piles) is produced by heating slowly two egg yolks in a split culm. The mixture, with the oil given off, is claimed to be wonderfully effective. In the same way, wine kept in green bamboo for a few days has, they say, a greatly improved flavor.

In both China and Japan the art of painting in black ink, so closely related to calligraphy that a written poem grows naturally from the picture, is largely based upon the brushwork necessary to limn bamboo. There are at least a dozen prescribed ways of using basic strokes to paint the twigs and leaves; each stroke, whose lightness suggests the motion of birds, possesses some such name as "startled crow" or "goose alighting."

In his own handwriting, on a painting of bamboo, there is a poem by the great Chinese emperor Ch'ien-lung in which he praises the plant's virtues. He says that the mere sight of it brings calm, and that the proper place for the scholar is in the country, where the uprightness of bamboo goes together with that of the gentleman.

Words are impermanent and may be of small value. This is true also of the modest bamboo. But bamboo is a medium that can hold their ephemerality. As the brush in calligraphy, now swift, now deliberate, bamboo confers on fleeting syllables the permanency of literature which can outlast civilizations.

From recording the transient to mastering the eternal, bamboo is the adaptable medium also that fashions jade—that precious substance treasured throughout the ages. The smooth and enduring surface of jade cannot be cut by metal: it may only be worked as the sea works, by abrasion. It is bamboo which is the vehicle for the cutting material in this timeless process. Fine minerals impregnate sticks and small wheels of the serviceable bamboo which revolve endlessly against the unyielding jade—to discover in time the artist's intention within, using the skill and patience that have forever signified the East.

Bamboo: Its Beauty and Uses

In Nature

To be among growing bamboo in a grove is to be surrounded by a sense of peace. This experience does not come from a dead stillness: the branches and the leaves far above are unresting, and the vitality of the young sprouts in spring is almost tangible. It is compounded of the silence, of the color—sunlight, delicately reticulated through foliage, shifting on the new green of the leaves—and the architectural regularity of the surrounding bamboo pillars, evoking recollections of quiet cathedrals. One can absorb in this mood the tapering strength of the great shafts which culminate overhead in a trembling plumage, and be gladdened also by their bright freshness and simplicity.

In winter, the sense of growth is less urgent and the communion becomes closer, reaching its distillation in the soft rustle made by snowflakes falling on bamboo in the depth of the night.

The pictorial aspect of bamboo and the night is reflected in a haiku by Kodo:

> *Moonlight slides up and down*
> *the stems of young bamboo*
> *swayed by the night breeze.*

Genesis: a young bamboo shoot

Thin black bamboo at the Seifu-so gardens, Kyoto

⇐ *Bamboo in flower*

Various types of bamboo

"Tortoise-shell" bamboo—a rare variation ⇨

Bamboo in section. The lower right-hand picture shows how few large roots a culm possesses.

A grove of black bamboo, Seifu-so gardens, Kyoto

Bamboo, cut and stacked

The production of square bamboo by growing it within frames, discussed on pages 206–8. The last picture shows "sesame-seed" bamboo on the left and chemically stained bamboo on the right.

Bending in a high wind

Looking down on the leaves of dwarf sasa bamboo

In Garden and House

Bamboo is completely appropriate to the garden. It takes a place there quietly, carrying out its tasks well but with no ostentation. As fence, broom, or trellis, or serving other ends, bamboo remains so near to its growing form that it is still entirely natural; it gives no feeling of something alien, imported to mar the harmony of a garden, that delicate simulation of a setting created by Nature alone. The manner in which objects of bamboo blend with the whole comes from the suggestion that Nature is aiding, rather than man imposing. Mellowing from green to gold with exposure, bamboo becomes more apparent to the eye but retains always its close kinship with growing things.

In large parts of the Far East, early houses were made entirely of bamboo. Many still are. This is not surprising: as a building material bamboo is cheap and abundant, easily worked, strong, and weather-resistant. These qualities have been combined to fashion an infinite variety of shelters for man. Pillar, beam ceiling, floor, walls of woven bamboo covered with clay, roofs from split semicircles facing up and down alternately—rustic or sophisticated, all its architectural uses reveal the beauty of simplicity. In the stern buildings of modern cities, bamboo has been outwardly displaced. But the ages have linked it so inseparably with the concept of home itself that bamboo still finds a place, greater or lesser, inside every house.

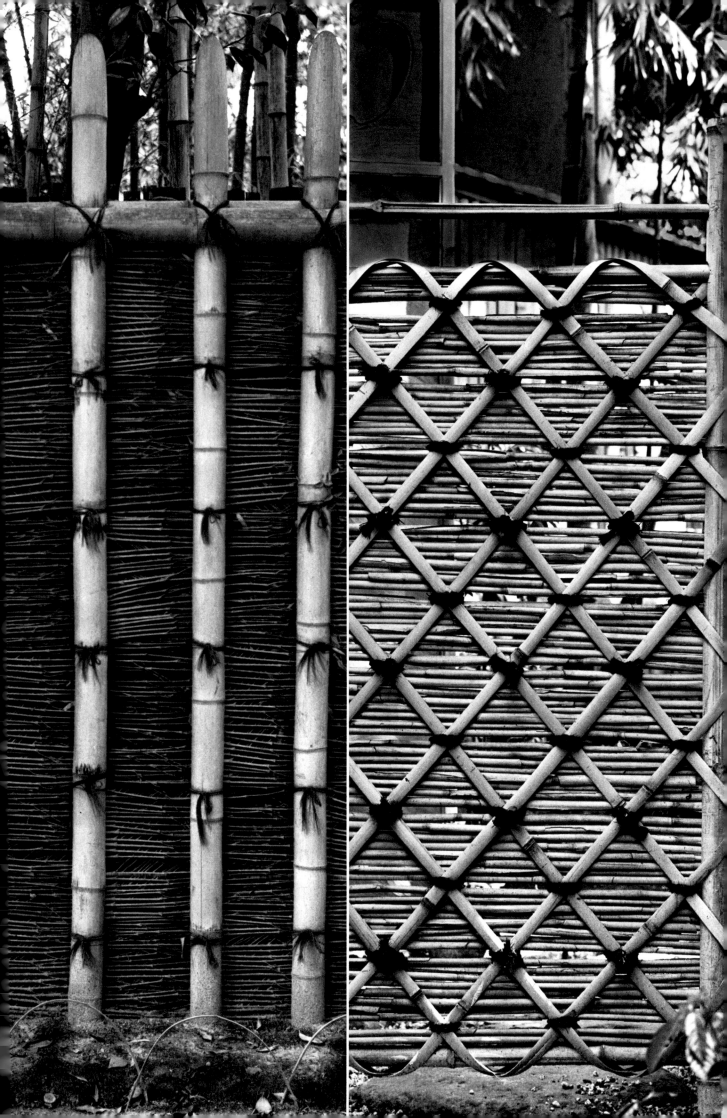

As shown on the preceding and following pages, the Japanese bamboo fence has many varieties. Traditional construction demands that only natural materials be used, coconut fibers serving to bind the cut bamboo together. Famous palaces and temples have preserved their own styles of fencing, to which their name is given.

The outer fence of the Katsura Detached Palace, Kyoto

Even fences in the same basic pattern can vary subtly in details as seen here in diagonal fences of two Kyoto temple gardens, Koetsu-ji (above) and Ryoan-ji (top of opposite page). Note how the angles of the diagonals and the contruction of the top rails differ. Also noteworthy is the graceful use of bamboo hoops along a path.

A simple fence sets off a teahouse in the private grounds of the Imperial Palace, Kyoto.

In the garden of the Kinkaku-ji (the Golden Pavilion), Kyoto

A fence of untrimmed bamboo at the riverside teahouse Shimbado, Kyoto

The studied simplicity of the entrance to a Japanese house

A path through the bamboo
grove of a Kyoto stroll garden

In the garden of the International Hotel, Kyoto

Stone lantern at the Shimbado, Kyoto

62.

The yarai *style of fence may be seen at the* Daitoku-ji, *one of Kyoto's largest temples*

A bamboo gate at the Tsuruya, one of Kyoto's beautiful garden restaurants

Part of the Imperial Palace, Kyoto. The guttering and rain-pipes are bamboo.

Bamboo ridges atop the thatched roof of an old-fashioned teahouse, Shinjuku Gardens, Tokyo

Bamboo "sleeve" screens sometimes project from a house into its garden to create nooks and also to give more privacy to adjoining rooms opening into the garden

⇐ *Slanting bamboo frames are sometimes used to protect exterior walls against damage by the elements or by animals*

Bamboo details in the Katsura Detached Palace, Kyoto. The platform above, built out over a small lake, was used for viewing the autumn moon.

Winter views of the Hotel New Otani's garden in Tokyo

Bamboo reflections in a pool at the Tsuruya

This tube, mounted on a central pivot, helps to scare away deer from gardens. It is filled with water from a pipe and, tipping its load, strikes the rock behind with a hollow report. It falls back empty and the cycle repeats itself.

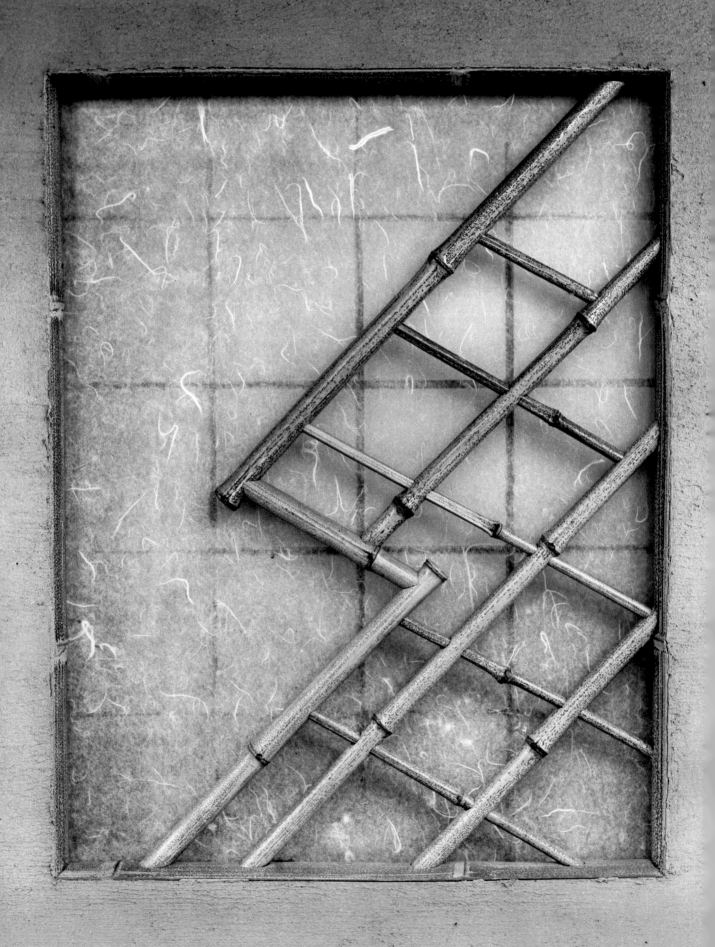

Window lattices of bamboo and reed, sometime laced with vine. Note the use of square bamboo in the bell-shaped window.

Bamboo is interlaced in many ways to make windows. It is also employed in ceilings and walls, as the following pages show. On page 85, bamboo sets off a tokonoma—the small alcove in the main room of a Japanese house. Here objects of beauty are displayed and guests are honored by being seated in the place nearest it.

At the Bamboo Restaurant in the Miyako Hotel, Kyoto

A paper window with simple bamboo bars breaks the severity of an inner wall of a house

In Everyday Use

For making simple things, bamboo cannot be bettered. In the East, it is found almost everywhere, in all sizes, and is immediately available for working with the basic tools—a knife and saw. As an aid in eating and drinking, bamboo may be used with a minimum of preparation. As a material it is almost universal, having qualities that are unsurpassed—strength, lightness, rigidity, and flexibility; it is also cheap and abundant.

Traditionally, bamboo constructions are held together with pegs or bindings of split bamboo rather than with nails or other metal fasteners: the properties of the material itself have imposed a discipline of form. But the recent advent of synthetic adhesives has changed this. Now, bamboo may easily be glued together; it can be laminated also and worked like wood. This is a mixed blessing. The bamboo industry depended formerly on the hand skills still natural to older people, but these are becoming harder and harder to find. Having taken a new direction, and having admitted machinery, it has enlarged its scope and gained new life. However, the rules imposed during the centuries by the material itself are now overridden and many articles appear which are unsuitable and meretricious. This may well be temporary: the traditions of the bamboo crafts have deep roots.

A bamboo ladder

A bamboo rake called a kumade *(bear's paw)*

The frame of a Japanese umbrella, covered with heavy oiled paper, is made entirely of lacquered bamboo. The hinges are made by splitting the spokes and anchoring them with thread.

Baskets for oshibori—the small towels (hot or cold according to season) offered to diners in restaurants

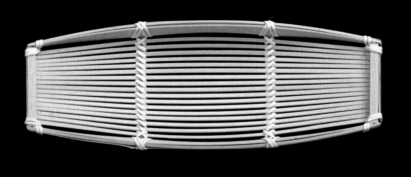
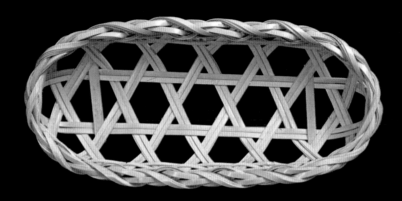
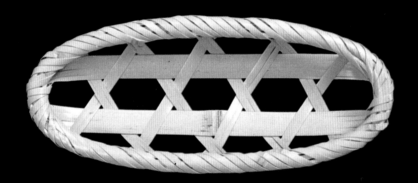
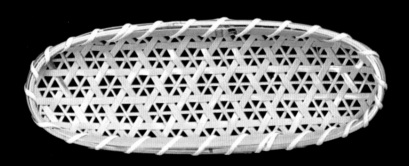
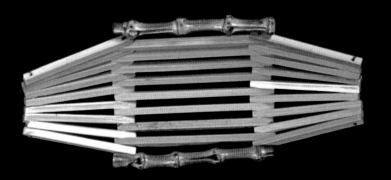

Plunged through the flexibly meshed sacks of rice, this rice-tester brings out samples for examination

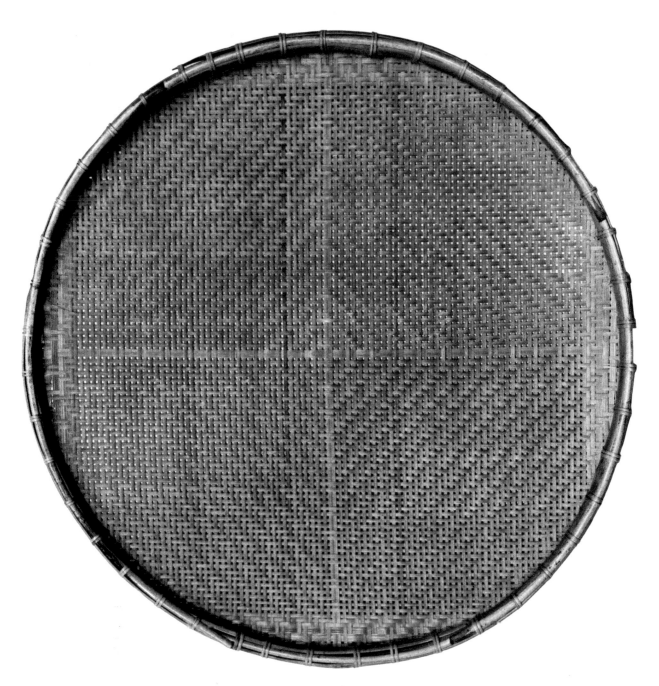

A winnow for grain

This brush—six inches wide—is used in China for painting or applying paste

Market baskets in Hong Kong and Cambodia

Fish tubs banded with bamboo hoops

A tea strainer

Throw-away baskets for vegetables

Basket weaving

An eel trap

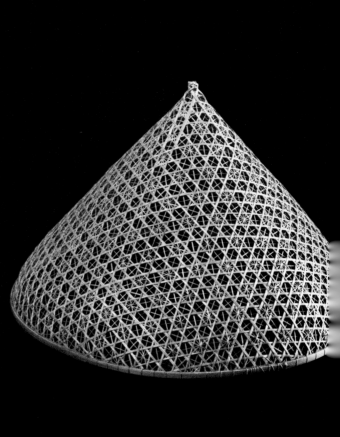

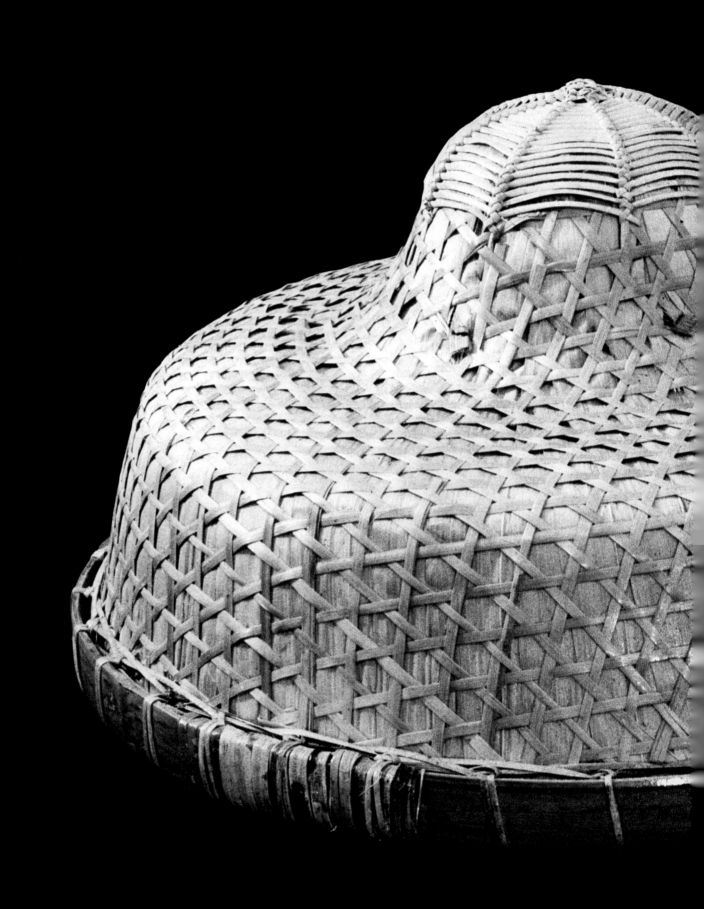

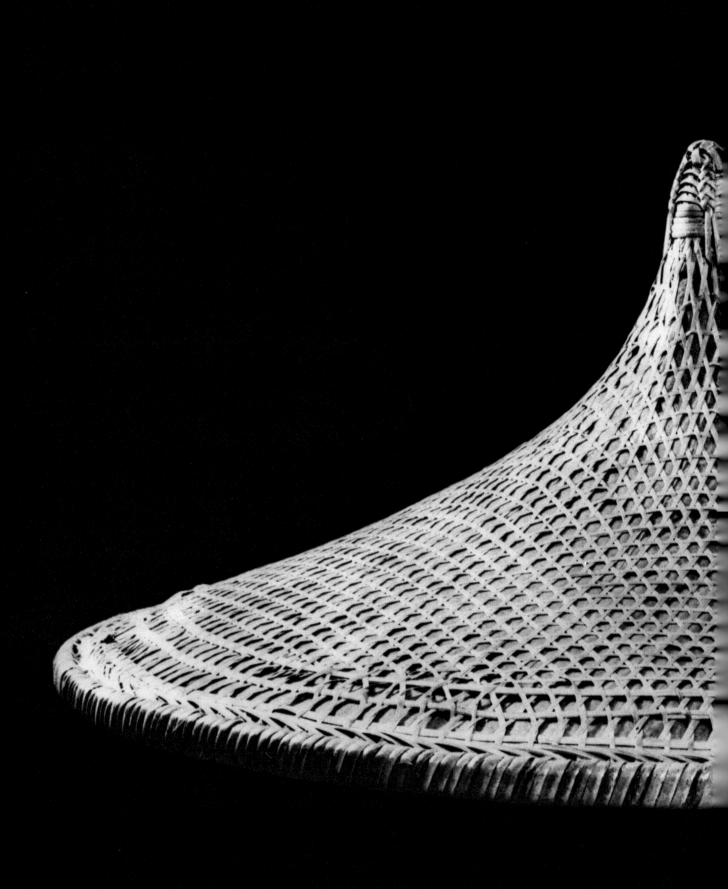

Sushi—patties of vinegar-flavored rice—served on bamboo

A bamboo shoot (called takenoko *or "bamboo's child") being prepared for eating*

Tea cakes on a bamboo plate

A Japanese lunch of rice and fish packed between two pieces of bamboo and then wrapped and tied in a bamboo sheath. The chopsticks are also held together with bamboo.

Bamboo spoons in various sizes

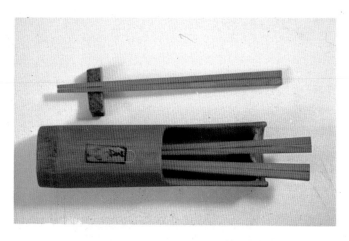

Bright-red rest and holder for chopsticks

A ladle for soba—the Japanese spaghetti

The first picture is of a carpenter's line. A thread is drawn from the reel across ink-soaked wadding in the recess and, released from tension, prints a straight line on the wood. At lower left, a simple fan from Bangkok, and on the right, a good-luck souvenir from a Japanese shrine, with imitation gold pieces and a bamboo rake for gathering them.

This bamboo toy represents a strolling shakuhachi player. Since they were often mendicant priests, they concealed their faces under baskets to show their unworldliness.

Above, a bamboo toy with movable head. Left, two writing brushes. The first is a simple length of bamboo, one end of which was buried in the ground for several months until bacterial action broke it down into rough filaments. The second, made of animal hair, has a bamboo handle with the type of brush and the maker's name carved into it.

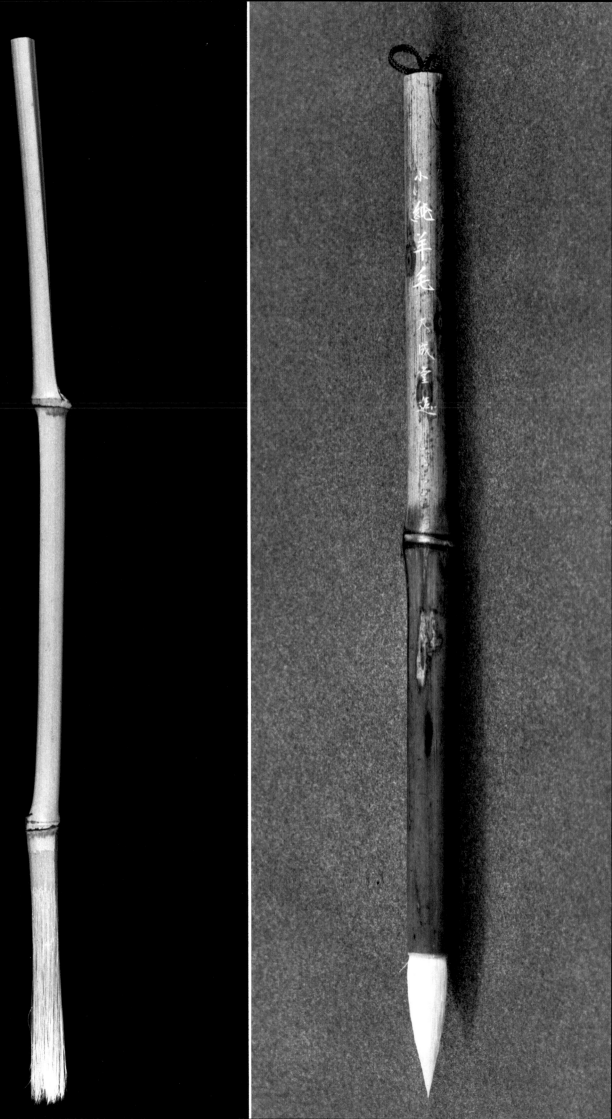

A collection of bamboo fishing poles

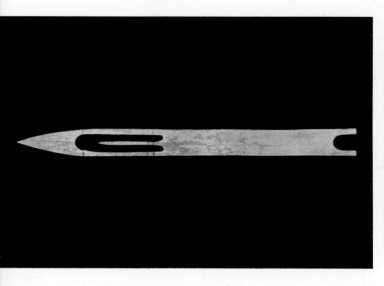

Net-mending needle

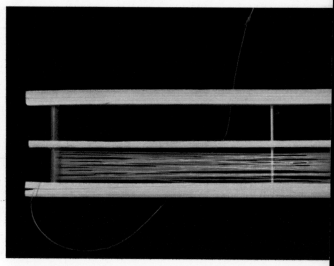

Fishing reel

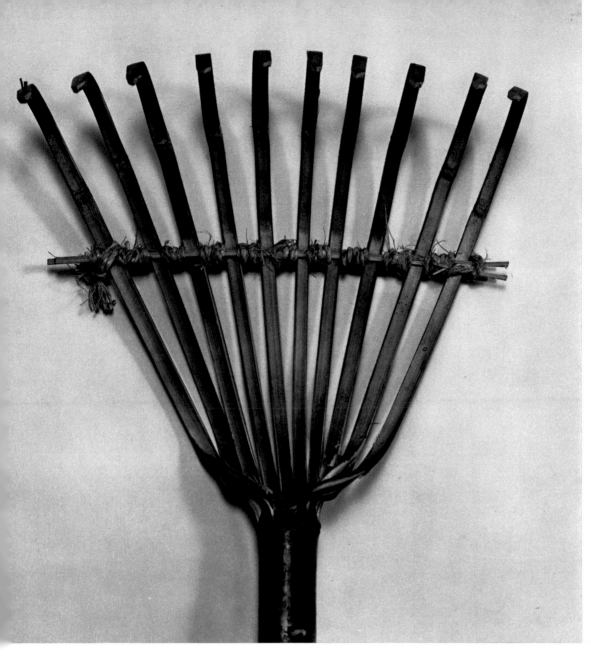

Bamboo rake

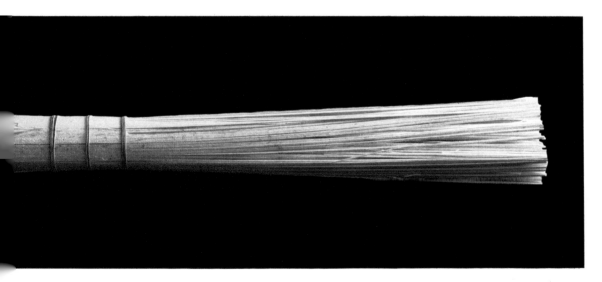

Dish scourer of split bamboo

Bamboo has excellent dimensional stability and is a superb material for slide rules and proportional scales—or (below) even clothespins.

(Opposite page) As seen in the upper pictures, split bamboo has been made into blinds for centuries. In the more expensive versions (left), the joints are arranged to produce patterns. The cheaper, commercial blind shown to the right is still in the weaving machine.

The lower pictures show a partition made of lengths of square bamboo (left) and a ceiling inlaid with bits of bamboo (right).

(Page 126) Bamboo is used as scaffolding all over the East, as in this Hong Kong scene. The intersections and joints are lashed with rattan, which is tied on wet and shrinks to take a very tight grip.

In Crafts and Art

The crafts of bamboo are very pure. This is because bamboo itself as a raw material has almost no financial value. Thus the estimation set on those articles made by masters of the traditional crafts derives entirely from the skill and art of their hands. It is the craftsmanship itself which is prized.

This is more true since bamboo is not by its nature a material which can be expected to last a long time. Nor, generally, are the things made from it of a kind to be kept carefully or mended when broken. In fact, bamboo articles older than a couple of centuries are rare indeed—except for the small collection deposited in A.D. 756 at Nara. This represents a group of everyday articles used by a Japanese emperor, which are revealing in the sophisticated detail brought to such simple objects.

The craftsmen who work bamboo finely are few now. They are individuals, the latest in a line of living tradition. In this they remain secure, respecting instinctively the material and their inherited craft—and the old tools also. These men work simply, with a patience that shows their complete mastery of the trade. They are artisans all, artists often, and from their modesty there always shines contentment.

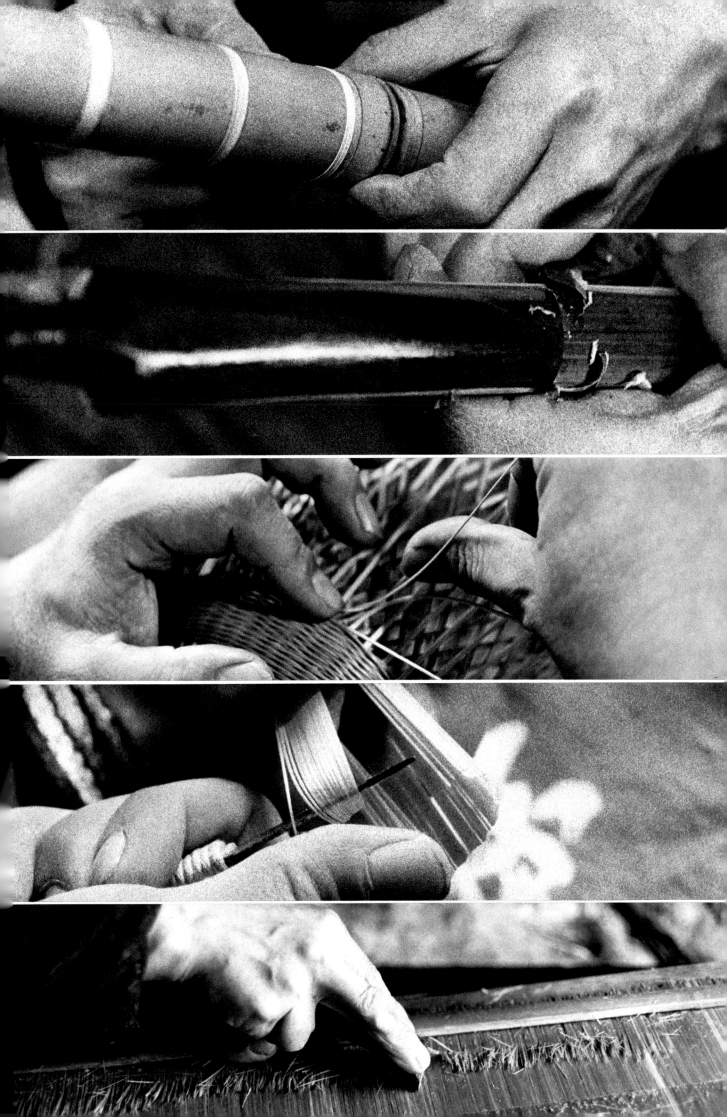

FANS *have been made for hundreds of years in Kyoto, the former capital. For elegance and an unhurried life, Kyoto still reigns supreme in Japan. Thus it is natural that the art of fan making, so long associated with refinement in manners, should flourish there still—as, indeed, does a great affection for bamboo.*

Fans are relatively expendable and their makers have always produced them in quantity. Traditionally, the operations have long been divided for the sake of economy in manufacture. There are over twenty-five steps in making a sophisticated fan and even today they are carried out in not less than seven different places, each of which has its own specialty. These are family groups, working deftly in small rooms, whose network of cooperation spreads over the city like the roots of bamboo itself.

In the heat of the East, the fan remains a great convenience. There are many variations on the two types—folding and rigid—and the colorful possibilities for their decoration are infinite. It is a long time since that strange symbol of the samurai, the cast-iron war fan, has been used; but in summer fans are still found everywhere and remain part of a Japanese gentleman's formal attire. Of late, however, fan-makers have been complaining that air conditioning is threatening their livelihood.

1. In preparing bamboo for fans, the joints are first cut out and the pieces sawn to the right length. 2. They are then split in half, to be divided further by this simple machine—or by hand. 3. The hard outer rind is peeled off. 4. Bundles ready for splitting. 5. About a thousand slips at a time are gathered into a block and marked with a point for shaping. 6. With a special scoop-shaped chisel—sharpened constantly—a curved section is removed. 7 and 8. The shoulders are cleaned out and the ends trimmed. 9. Semifinished slips, already drilled, standing among bamboo shavings.

10

11

14

15

10. *Fan skeletons awaiting covering. These fans, destined for expensive teahouses, have the upper part of the ribs reduced to the thinness of paper.* 11. *The hand-painted, pleated body of the fan, made of two layers of paper.* 12. *Opening rib pockets.* 13. *By blowing, the pockets are opened wide.* 14. *The ribs are pasted.* 15. *The pasted ribs are inserted, one by one, with great speed and dexterity.* 16. *Finished fans drying under slight pressure, after which they will be perfumed and packaged.*

12

13

16

For the second shot,
Rattling his stiff paper robes
He turns his shoulder to
The festive target, that is nothing,
With ceremonious ease
Lays fingers on the string
And lifts the bow above his head.

Moving his hands apart,
He brings the arrow to his level eye,
In one gesture, or lack of it,
Indifferent, formal, bold,
Lets fly the shaft that is
Himself, and splits
The first arrow at the centre of the gold.

MAKING BOWS AND ARROWS *is a craft of long continuity, which was shown by Hambei Hazu VIII when he adopted the name of his teacher. The bows his school has been making in Kyoto for three hundred years are recognizable by their special shape. This shaping is the most difficult part of making a bow and takes at least ten years to master. Since these particular bows are primarily intended to be used on horseback, their point of balance is considerably below center.*

Bamboo for bows is usually yellow and is left standing for two or three years in order to attain this color. After the oil is removed, the bamboo is cut and laminated to wood, and the bow is left under tension for three months to take its approximate shape. The finishing demands exact judgment of balance and curve. The bowmaker must scrape and test to bring the shape to a perfection which varies individually and cannot be set down.

In making arrows, the feathers are of great importance. They come from the hawk's tail and, for the best, only two can be used. In a set of four arrows, twelve matching feathers are necessary and for the finest shafts no trimming or adjustment is permitted.

The cult of archery is still strong in Japan. More than in most other sports, rigorous training is combined with such a discipline of thought that it is believed the experienced archer need not look at his target; he will strike it if he is only thinking about it correctly. This feeling is captured in the accompanying poem by James Kirkup, called "Zen Archer" (from Refusal to Conform, *quoted by permission of Oxford University Press).*

THE BOW

1. The bow is built up by laminating together several strips of bamboo and wood (catalpa, Japanese oak, and mulberry are favored varieties), which are finally covered with two outer strips of bamboo showing their natural surface. Before the strips are glued together, the maker's signature is inked on the innermost one—never to be seen again.
2. After gluing, the bow is bound with a cord. It is tightened and shaped by eye with bamboo wedges to give the bas

...rves, then left to set for three months. (Inset: section showing some of the laminated strips.) 3. The gauges and ...ols of the craft. 4. Thickness is checked with the wooden gauge that has been used daily in the bowmaker's family for ...er a century. 5 and 6. Trimming and scraping the edge. 7. The curve and strength are checked. (On the right is ...e bending horse.) 8. Fitting the wooden ends.

9

10

11

9 and 10. *A trial bowstring is put on to verify the balance. The bowstrings are made of hemp solidified with resin. Each has a distinct and different musical note. 11 and 12. The final curves are set into the bow with the bending horse which has been in use for many generations. Thus the bow has reached its basic shape. Later, various finishing touches are added—lacquer, bindings of rattan, a leather handgrip, and a hemp bowstring.*

12

ARROWS

1. Arrows are made in sets of four from a very straight bamboo. The shafts are carefully matched for balance and weight. 2. Any crookedness is corrected by bending with a wooden stick after the shaft has been heated quickly over a brazier, after which the nodes are pared and trimmed. 4 and 5. The metal head is applied and forced home by striking the arrow into a log. 5. The nock, made of horn, is glued on and shaved flush by rotating the arrow against a knife. 6. The flights are cut from feathers and glued to the shaft. 7 and 8. The head and tail of the arrow are bound with twine. When the glue is dry, the arrow is finished.

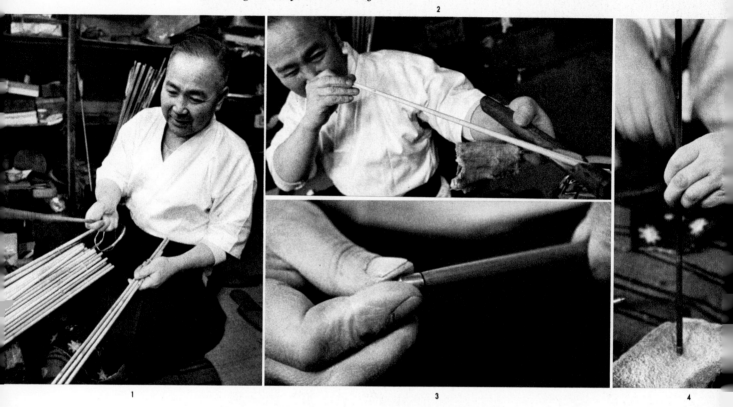

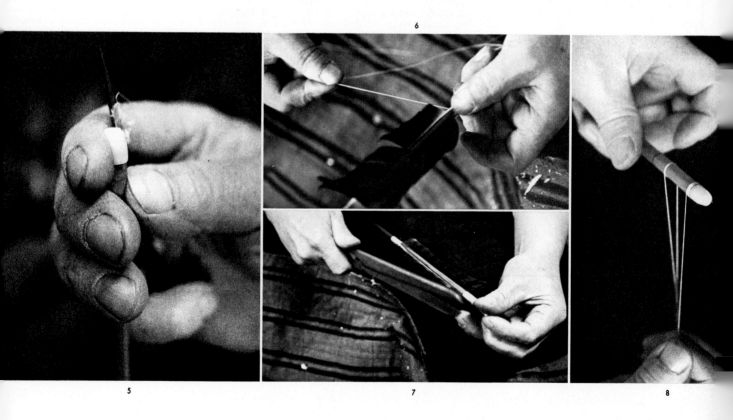

THE SHAKUHACHI, *Japan's vertical bamboo flute, acquired its name from the fact that, in its most popular form, and measured in the old Japanese units, its length is exactly one* shaku *and eight* (hachi) sun, *almost twenty-one and a half inches. The origins of this simple five-hole instrument go back at least to the seventh century in China. It produces the pentatonic scale and has a sound of great beauty and sadness. Many years of practice are necessary to master the subtle techniques needed for playing the shakuhachi.*

This maker, Kozo Kitahara, learned the trade from his father and continues it in his old studio. He is one of the fifteen who are left in Japan. He has been making shakuhachi for twenty-five years, even though he is still young. He brings to his work a sense of humbleness and dedication which is immediately communicated. Loving the bamboo itself, he cannot even bring himself to eat its shoots. In the winter, he goes up alone into the mountains, where he selects each one of the stems he will work. To stand thus in silence in the snow of a bamboo grove is his only holiday. The bamboo dries for three months on the roof of his house and then seasons in the dark for three years before its condition is ideal. During this seasoning process the stems are trimmed into rough lengths, leaving a portion of flaring root at the base of each, which will become the bell end of the finished instrument.

1

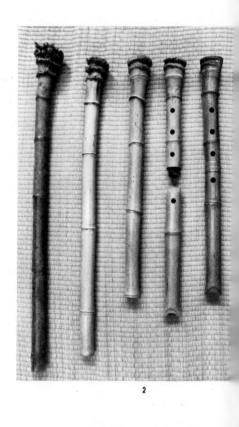

2

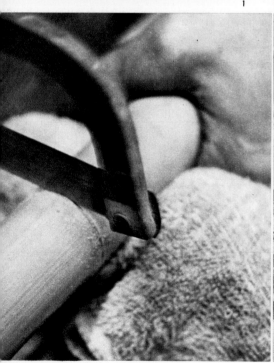

6

7

8

1. A stack of bamboo pieces which have been dried and seasoned for three years and are ready to be made into shakuhachi. 2. Shakuhachi in various stages of making. 3. The bamboo is heated and drops of oil which appear are wiped off. As this is done, the bamboo rapidly becomes whiter in color. 4. The culm is straightened by heating and bending under tension. 5. Being marked for cutting. 6 and 7. The center section is cut out and discarded. Only the ends of the piece are used. 8 and 9. To strengthen the tube and avoid splitting, bindings of twine are inserted in shallow grooves cut at either side of the joint and sometimes at other points as well; these are later strengthened still more by being covered with some other material such as rattan, shell, lacquer, or other decorative material. 10. Inspection.

11. Cutting the mouthpiece. 12. A piece of water-buffalo horn is glued in, cut off flush, and smoothed. 13. Sharkskin is used to secure a fine finish. 14. The finished mouthpiece. The tube of the instrument is open at both ends; the sound is produced by blowing down over the thin horn lip. 15. One of the many times the shakuhachi is smoothed and polished during manufacture. 16. Locating the hole at the base. 17 and 18. Drilling, finishing, and checking the bore of the instrument. 19. Deciding where to open the fingerholes. The distance between their centers is exactly one-tenth of the length. 20. Assessing the quality of the sound for the first time. Adjustments are made by cutting and lacquering until the tone is judged perfect.

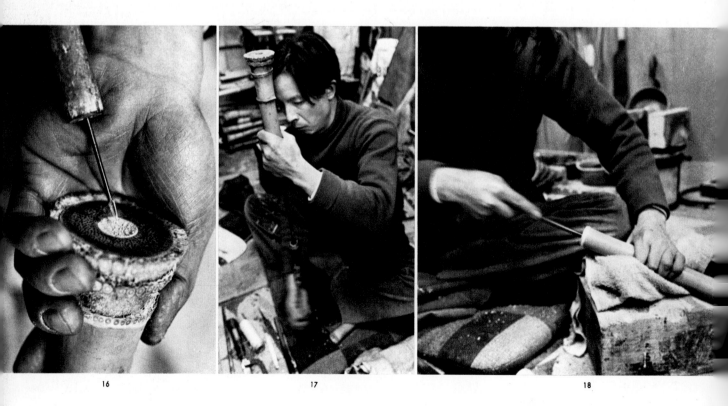

15

19

20

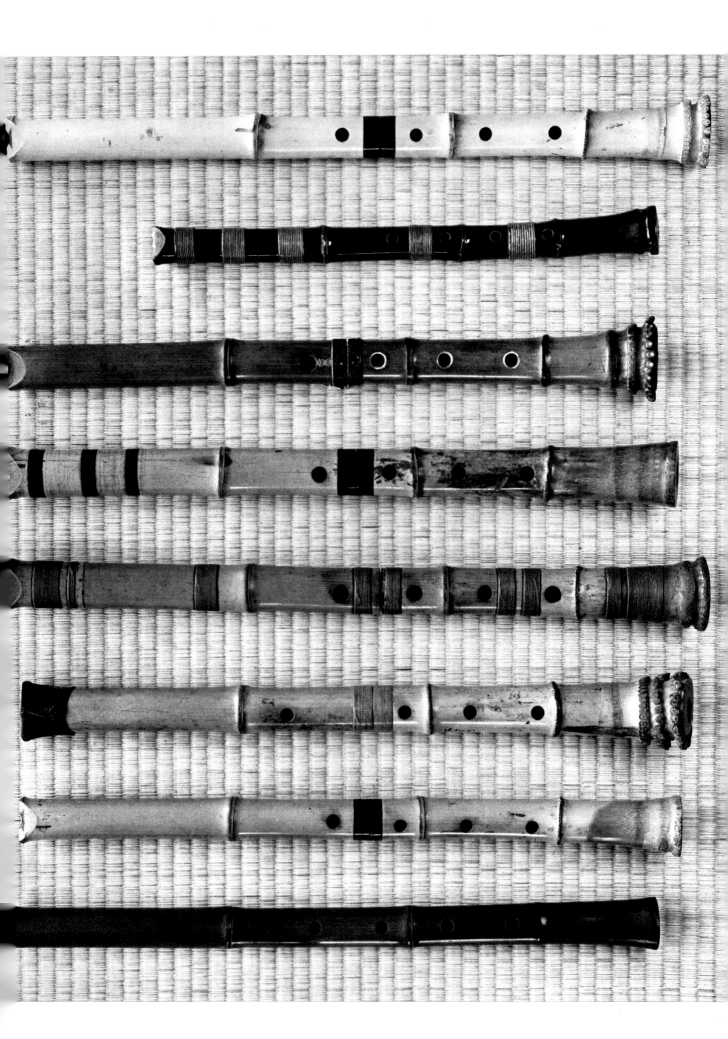

151

THE CHASEN *is peculiar to Japan and the Japanese tea ceremony. The word is translated as "tea whisk," since the instrument is used, with much the same motions as an egg whisk, to mix thoroughly in hot water the finely powdered green tea of the tea ceremony.*

The number of splines in the head can vary considerably, from 32 to 120, and there are names for many of the types. The more delicate are used for finer tea. Despite the skill involved in making the chasen, they are available at modest prices. As there are various schools of the tea ceremony, so there are varied types of bamboo from which the preferred chasen are made—smoked, white, spotted, or sometimes green bamboo.

The craftsman shown on the following pages is Sajiro Kubo, the eighth in a direct line to work in the trade. He makes the chasen together with his wife and father. There remain about a dozen makers in Japan, mostly grouped near Kyoto.

1. After wrapping with a temporary binding and immersing the bamboo in hot water for a few seconds, the first splits are made. (Note also the photographs below showing various stages of the making.) 2. Some of the outer rind is removed. 3 and 4. Sixteen radial splits are made and the splines are next split concentrically, making an inner and an outer ring. 5. The inner splines are broken out and the interior smoothed. 6. By bending the splines backward, the correct curve is put into base of the splines. 7. The splines are finally divided by a special knife. In good chasen

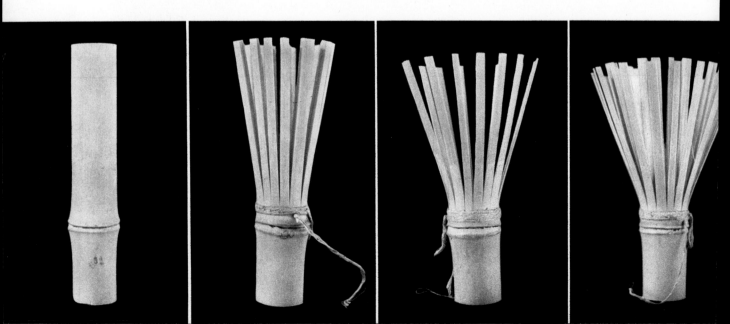

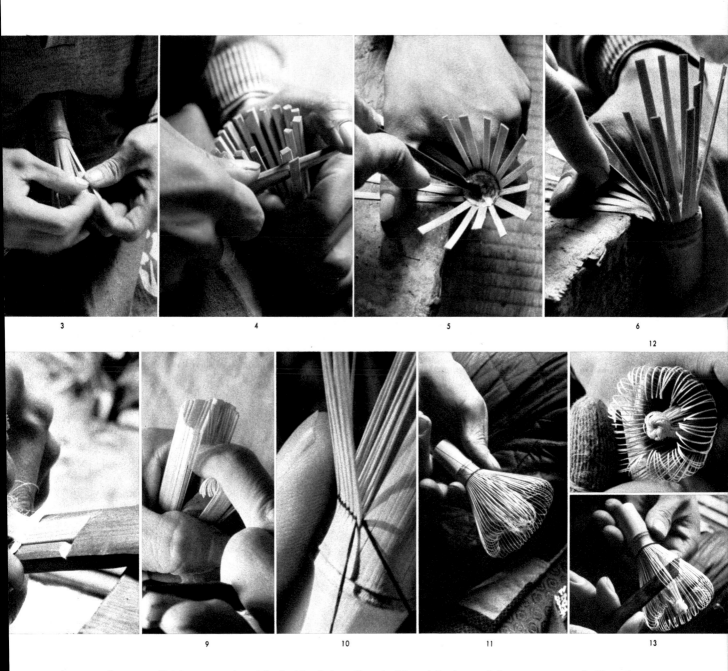

there are about 120 divisions. 8 and 9. The inside of the splines is thinned further and the tops are curved. To make this curving easier, the chasen is held briefly in steam. 10. A thread is passed alternately between the splines to separate them into inner and outer parts. This stringing is usually done by women. 11. The fully strung chasen. 12 and 13. Final adjustments are made to produce the perfect shape. During this part of the finishing a small sock is worn on the thumb.

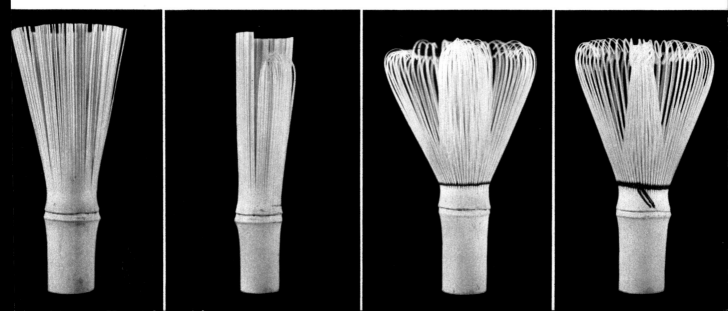

Tea-ceremony utensils shown on these two pages: *1*. Chashaku *(bamboo scoop for powdered tea)*, together with its tubular container and the top of its outer box. *2*. Kama *(iron kettle) and* hishaku *(bamboo water ladle)*. *3*. Chasen *(tea whisk)*, chashaku, *and* chawan *(bowl in which the tea is mixed and from which it is then drunk)*. *4*. Woman folding the fukusa *(ceremonial wiping cloth)*. *5*. Another view of the graceful hishaku.

The Japanese tea ceremony is intrinsically a simple thing. The refinements that have arisen within its framework are numerous and baffling to the Western mind. However, the ritual, which began as a movement away from artifice, derives from the simple act of making tea and offering it to a guest. Its prescribed utensils are also simple and made of common materials—bamboo, iron, pottery, and the like.

BASKET MAKING *is one of the oldest and, even today, most frequently seen crafts in which bamboo is widely used. Its products come in many different shapes for many different uses, ranging from the tiny baskets to hold toothpicks, to an enormous latticework that could easily hold two men and serves for flower arrangements in a hotel. But, whatever the size, the method of working remains the same. Bamboo is split into strips, more or less flexible according to their thickness and width, and these are woven into the finished design.*

It has happened in this way for thousands of years. Apart from a knife for splitting, or occasionally a simple pattern, there are no tools to speak of. The relationship between weaver and material is very close; but the distance between such a master as Iizuka and the unknown workers who make simple baskets for markets is very large, as between a great artist and a journeyman. But there is, all the same, a kinship which is bred of the same materials, techniques, and common utility. The distance sometimes comes from the time available for the task, time whereby the everyday basket may be raised to a point of art.

From the square agricultural trug or the strong round baskets on the backs of the gatherers of seaweed to the tracery of summer handbags or the delicate cages the Thai women balance on their long hair as support for their hats, the possibilities of bamboo in its woven form appear infinite. The weaving may also be two-dimensional, as in plates, screens, and so forth.

With the exception of the baskets on pages 165 and 169, all the baskets shown here (through page 172) were made by Shokansai Iizuka, one of Japan's most famous craftsmen in bamboo. As might be expected in a country where flower arrangement is considered a high art in itself, many of his baskets are intended for holding flowers. Such baskets are classified in much the same way as styles of flower arrangement into three categories—formal, semiformal, and informal—each of which has a number of strict rules. Iizuka is too great an artist to follow such rules blindly, but he is quite aware of them and their influence can be seen in his work.

1. Making the first split. 2. Fine division—the bamboo is drawn between two knives stuck into a block. 3, 4, and 5. Weaving.

A receptacle for small precious objects

OTHER BAMBOO CREATIONS

Box carved from bamboo

A carved box to hold seals

A small cup carved from bamboo

Brush holder for the writing table

A simple vase made by Shokado Shojo, a seventeenth-century flower-arrangement master

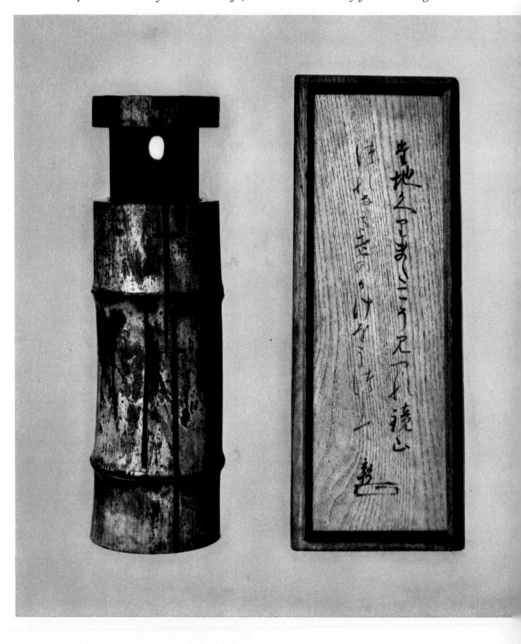

A Chinese brush holder, about 350 years old. The carving is deep and heavily undercut.

An elegant bamboo cane decorated with a silver frog

(Opposite page) An illustrated children's book from Bali

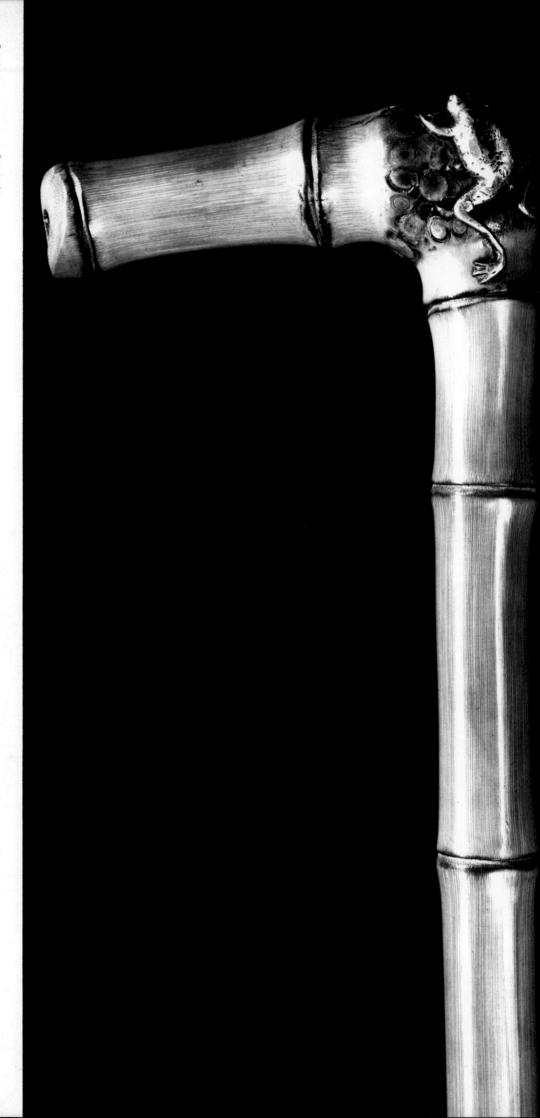

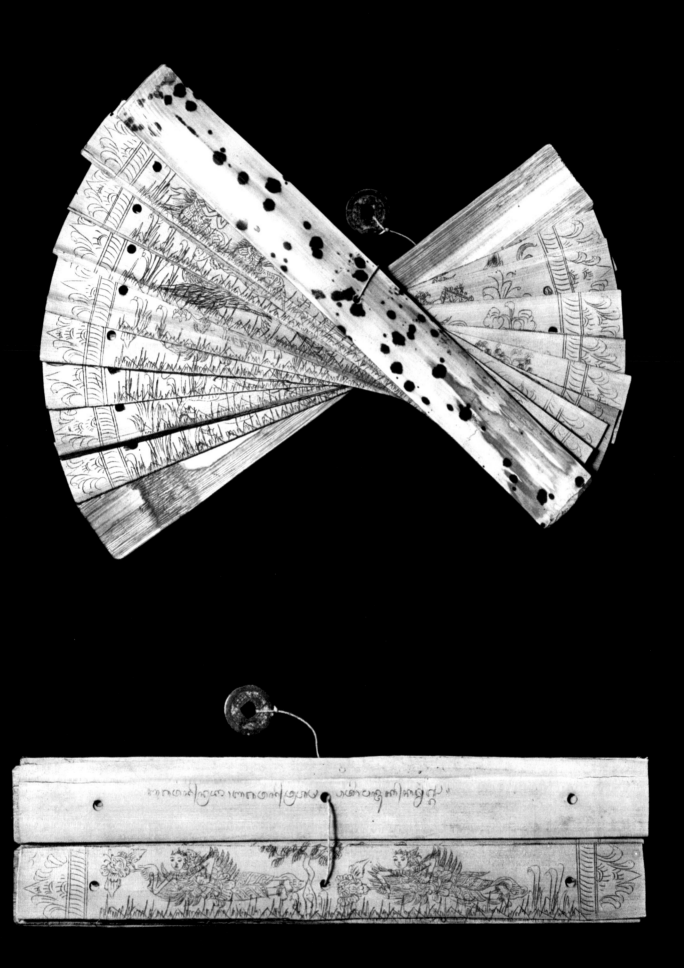

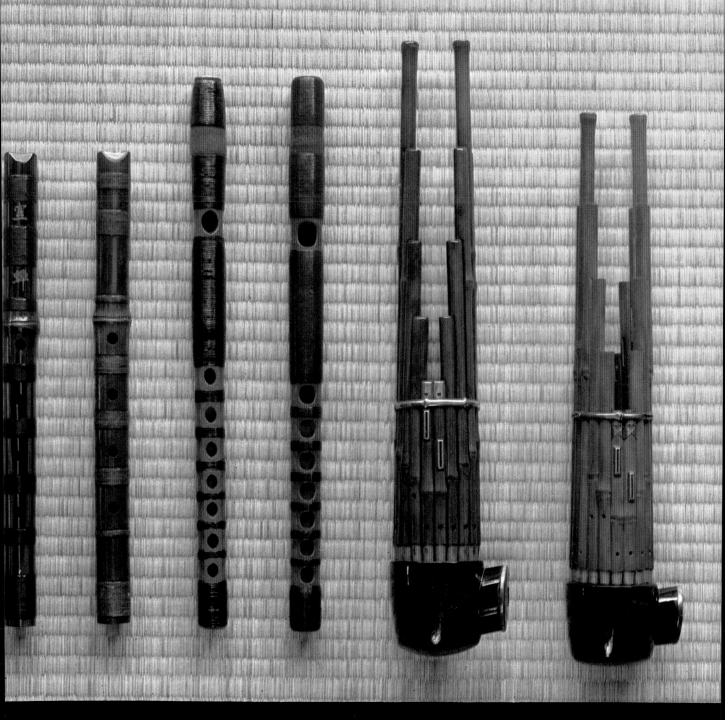
Japanese vertical flutes, transverse flutes, and sho—*small organs which are blown by mouth*

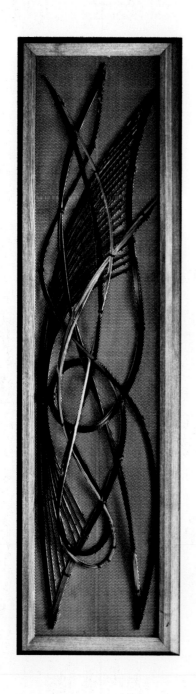  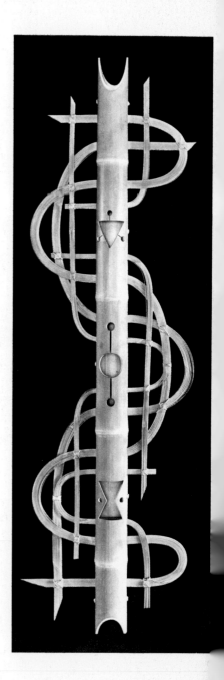

Contemporary wall plaques by Iizuka

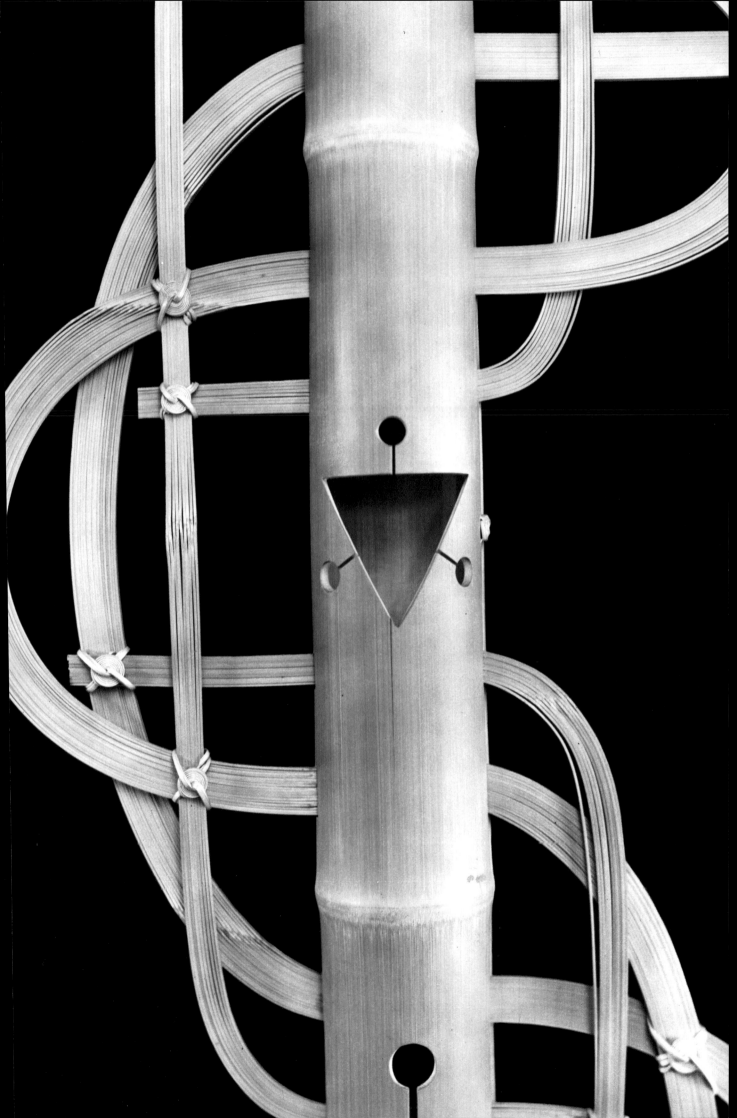

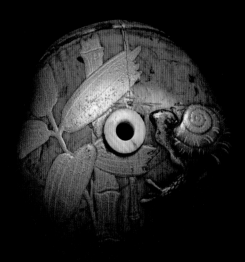

A netsuke made from lacquered bamboo. Netsuke were the decorated objects men wore on one end of the cord by which a small box for writing materials or medicine called an inro was suspended from the formal Japanese belt.

On the right, ivory carved to simulate bamboo. Insects on the reverse side (see below) were also carved from solid ivory.

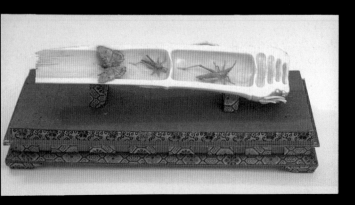

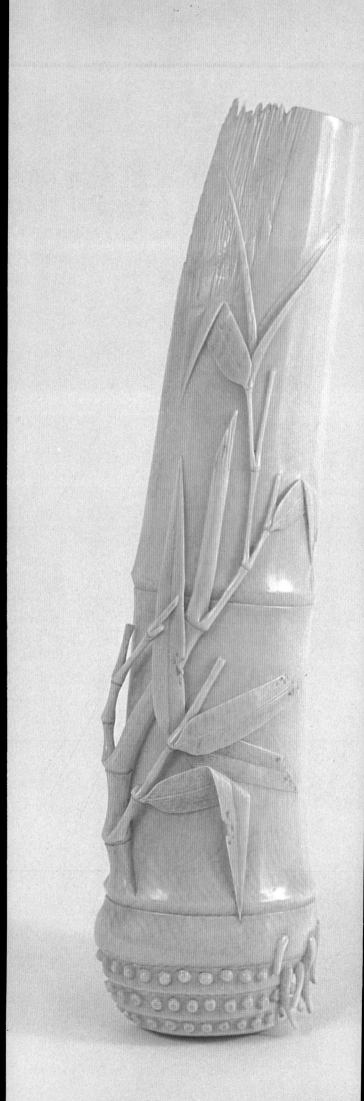

A portable tablet used for grinding ink

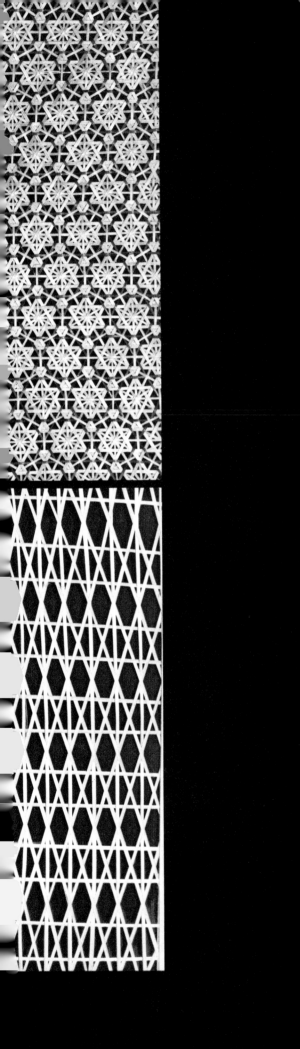

⇐ *Traditional patterns of bamboo weaving*

An inro of bamboo with small drawers (natural size)

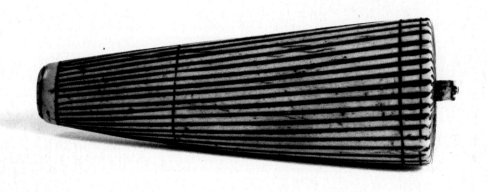

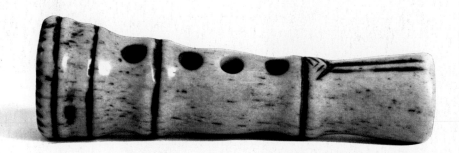

Netsuke, a little larger than natural size, representing bamboo objects— a bamboo shoot, an umbrella, and a shakuhachi

A contemporary bamboo design on a lady's kimono, in two colors of silver thread. Such one-of-a-kind kimono are very expensive.

One of the Hundred Views of Mount Fuji by Hokusai

Chinese ink painting on silk by Li K'an, about A.D. 1300

Bamboo: Its Growth and Cultivation

GROWTH

The speed of bamboo's growth is one of the wonders of nature. From the time when a sprout first comes through the ground to the completion of its growing, only about 60 days elapse. And after this period, the bamboo culm does little: it gains no height, does not thicken, and it undergoes only slight change. It is alive, of course, indeed flourishing, and seeks sustenance avidly—though apparently not for itself but to provide food for the sprouts that succeed it and for the network of rhizomes and young culms that constitute its family.

The vigor and fertility of bamboo is also remarkable. Every year there appear shoots in great number, which in turn produce more, and so it spreads until checked or disciplined by man. Bamboo propagates by the branching of its underground rhizome and does this asexually.

There are two main types: monopodial and sympodial. The first has a long underground rhizome with a single bud capable of development each year on the alternate side of each node or joint. While less than 10 percent of the buds germinate, and a smaller proportion fail to grow to full size, the yield is still large. The sprouts emerge from the ground well distanced from each other; they grow into culms singly and erect, the height before branches start being generally considerable.

The second type, clump-forming bamboo, has the new shoots coming from the parent plant itself. Here, the culm and the rhizome are one. The upper part of the short rhizome has buds, one of which develops into another very short rhizome and then turns upward to emerge from the ground as a secondary culm close to the parent. This process is repeated yearly, with the result that all the culms are grouped closely together. The branchless section of culm in the sympodial species is generally short.

Even in the same grove, there can be a difference of up to 2 months in the times at which sprouts appear from the soil. In general, the earlier sprouts are the best, developing into larger culms of higher quality. Bamboo in Japan is mainly monopodial, its main season for growth is March to June; shoots of moso (*Phyllostachys pubescens*) appear mostly in the first 3 weeks of April.

Monopodial bamboo does practically all its growing in a single month. To be stricter, the very top part—about 7 percent of the total length —goes on developing for a further month, but its gain in height being slow and slight, the only appreciable development is that of the culm. After this period there is no further extension.

Sympodial bamboo is slower in development, with no clear distinction between the two periods of growth. It takes from 80 to 120 days to reach its full height. For both types, the growing period is longer in the case of the earliest sprouts.

As observed, the rate of bamboo's growth is phenomenal, faster than that of any other plant. For example, during a single day, the Japanese authority Koichiro Ueda recorded that moso grew 46.8 inches while madake (*Phyllostachys bambusoides*) grew 47.6 inches. (These observations were made in Kyoto Prefecture in 1955 and 1956.)

The single rhizome of monopodial bamboo

also reaches its full length in one season. It lives for about 10 years. However, every node of a rhizome possesses buds, one of which may develop each year into either a culm or another rhizome—but generally the bud on the leading node turns into a rhizome, and it is in this manner that the underground system extends itself. The yearly growth ranges from 3 to 10 feet and sometimes up to 20. The complexity of the underground network of living and dead rhizomes is formidable, as is its length and weight. The figures in Table 1 demonstrating this were established in 1955 in Kyoto by the excavation of a quarter-acre of a grove.

The number of new culms in a grove fluctuates each year: normally, a good year alternates with one of lower production. In the poor years there is a smaller number of culms of inferior quality. The amount of bamboo produced may vary between wide limits, which depend on weather, soil, management, and other conditions. In a grove of madake, the annual total of new culms per acre is 400 to 1,500 while moso will yield 200 to 600. The yearly number of new culms in a good grove is *less* than in a poor grove, where the culms are more numerous but smaller.

The age of the rhizome is very relevant in its effect on which buds germinate and which do not. The 3- to 5-year-old rhizomes are the most fertile in producing new culms, while sprouts are rare from rhizomes over 10 years old (see Figure 1).

Bamboo culms grown even in the same year and in the same grove differ in size. Generally, large-size culms develop from the thicker, younger rhizomes. But even the sizes of the culms from the same rhizome can vary, depending on the amount of nutrients in the soil.

It must be remembered that the culm grows so rapidly that it has no time to provide sustenance for itself and therefore its growth is completely dependent upon the nourishment it receives through the rhizome and the "mother

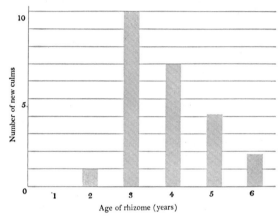

FIG. 1. RELATIONSHIP BETWEEN RHIZOME AGE AND NUMBER OF CULMS

bamboo." Its quality is thus affected considerably by the state of the mother plant. This can be convincingly demonstrated. Table 2 shows, first, the increase in size of new growth the year after fertilizing a typical madake grove; and second, in a different grove, the decrease, resulting from the drastic reduction of nutrients reaching the rhizomes, the year after cutting all the culms.

TABLE 1. LENGTH AND WEIGHT OF RHIZOMES

SPECIES	TOTAL LENGTH (yards)	FRESH WEIGHT (pounds)
Phyllostachys pubescens	2,300–10,400	3,300–14,800
Ph. bambusoides	5,800–17,000	4,400–19,000
Ph. nigra	13,000–16,500	2,200– 6,600
Pleioblastus pubescens	43,000–53,000	900– 3,300

The difference in the sizes of new culms can be adjusted by attention; it also appears that the characteristic of bamboo to grow well one year and badly the next may be controlled by fertilizing or by other means.

There is a close relationship between the diameter and the height of a culm, expressed graphically in Figure 2. Note that it is a custom to measure the thickness of a culm, either its diameter or circumference at eye level. As a rough rule-of-thumb it can be said that the height of a moso culm will be about 40 times its girth, while this factor increases to about 60 in the case of madake.

The weight of a culm at cutting is designated its fresh weight; drying reduces this figure by about 30 percent and gives the air-dry weight. Bamboo is sold in Japan by the *soku*, which is the unit of volume for culms gathered into a bundle about a foot in circumference. The fresh weight of one *soku* is approximately 70 to 75 pounds.

The leaves of bamboo fall yearly, but they give place immediately to new ones. The monopodial species change foliage in spring whereas the sympodial plants shed theirs in winter. Bamboo, like any other plant, gains much nutrition from its leaves: those species with a larger leaf area absorb more water and grow with more vigor. Moso has several times more leaves than madake, the weights being roughly 4 to 6 compared with 2 to 4 tons per acre. It is interesting to note that the weight of the leaves in a grove is roughly equal to the weight of the annual new culms.

When the conditions of its environment become unfavorable, the bamboo seeks security by developing larger leaves. This is particularly

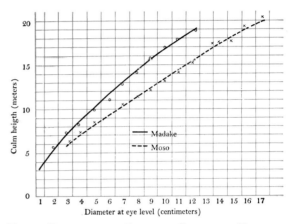

FIG. 2. RELATIONSHIP BETWEEN GIRTH AND HEIGHT

noticeable after the rare flowering of bamboo, when the small regenerated shoots that do appear bear for a long while leaves which are especially large.

PLANTING AND CULTIVATION

Bamboo, being of such vigorous growth, will germinate from almost any young rhizome or

TABLE 2. EFFECT OF FERTILIZING AND CLEAR-FELLING ON RHIZOME AND CULM SIZE

PH. BAMBUSOIDES	BEFORE FERTILIZING	AFTER FERTILIZING	BEFORE CLEAR-FELLING	AFTER CLEAR-FELLING
a) Diameter of rhizome in inches	0.55	0.55	0.95	0.87
b) Diameter of culm in inches	0.59	0.78	2.52	0.63
Ratio of b divided by a	1.10	1.40	2.70	0.70
In addition, the number of new culms increased after fertilizing from 83 to 152 in an area of about 100 square yards.				

cutting, but there are certain preferred methods of planting as shown in Figure 3. A few comments on these methods as numbered in the sketches:

1) The monopodial culm chosen must be young, having sprouted in the same or previous year. The top part should be cut off, but several of the small lower branches should be left. The total length of the trimmed culm should be about 5 feet.

The rhizome is very important in all monopodial plantings. It should be yellowish in color and have good buds. The best length is 15 to 40 inches with over 10 nodes and buds. It should be cut carefully with a saw and not an axe to avoid undue shock to the young growth. The rhizome should be planted to a depth of 1 foot.

2) Here the culm is trimmed shorter, to a length of about 1 foot.

3) A section of strong monopodial rhizome 2 or 3 years old with plenty of small fibrous roots can also be used. It should be about 20 inches in length with 10 to 15 nodes. This method of planting is recommended for transport to a distance: the rhizome should be wrapped in damp moss, then plastic sheeting, after the soil has been washed off. Cut rhizomes handled in this manner are usually first planted in the nursery about 8 inches deep, then transplanted the following spring when the new culms start to shoot.

It should be noted that monopodial bamboo takes root only with difficulty from a culm without a rhizome.

4) The sympodial type of bamboo having no long rhizome, a 1-year culm is dug up and cut off under the ground as near as possible to the mother bamboo. The culm itself may be 1 to 5 feet long. If the culm is over 2 years old, it may be incapable of propagation.

5) A length of culm of the sympodial type 20 to 40 inches long with 3 or 4 nodes roots easily. It should be branchless and 1 or 2 years old. It is buried entirely in the soil to a depth of 1 foot.

6) An alternative to the foregoing method is to plant the culm at a slight angle with one-third of its length above ground.

In addition, though it happens rarely, the sasa species may develop seed, as do some species of the sympodial type. Seeds are sown in a nursery, but the newly grown shoots are slender for several years. Since culms of a normal size are not produced for about 10 seasons, other methods are usually selected for the replanting of bamboo groves.

Bamboo can be rooted at any time of the year by careful treatment. However, the best months for replanting the monopodial species are October, November, and December. For sympodial

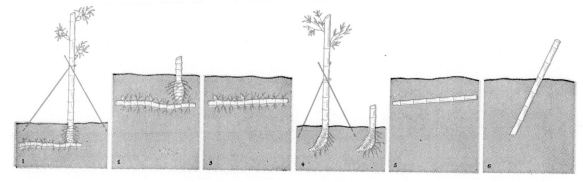

Fig. 3. Preferred Methods of Planting Bamboo. Monopodial: 1) Longer length of culm with rhizome attached. 2) Shorter length of culm with rhizome attached. 3) A length of rhizome alone. Sympodial: 4) Longer or shorter lenghts of culm and rhizome. 5) Horizontal length of culm. 6) Slanted length of culm.

bamboos, the rainy period in the spring and early summer is the optimum season. The suitable shoots should be selected when the rhizome buds show signs of slight swelling; replanting should be carried out reasonably soon after digging up the culm and rhizome in order that the rootlets will not dry out.

Generally, 125 to 200 cuttings per acre is the correct density for planting a grove of monopodial bamboo.

Bamboo that has been freshly transplanted should be watered from time to time. At the beginning this should be done at least weekly, but not more than every 4 to 5 days. Fertilizers need to be applied in spring and summer to assist growth. On sloping land the rhizomes should be planted uphill of the culms. Trees are desirable around the grove to act as a windbreak to the young plants and also to give shade.

Bamboo and Soil Erosion. Bamboo is valuable in controlling soil erosion in areas that are poorly adapted for other crops. It grows well both on steep hillsides and along the banks of rivers, its most important features being the interlocking root system, the mulch it produces from its leaves, and its habits of propagation without attention. The sympodial types are best suited for this purpose.

The amount of small roots in a clump of *Bambusa tulda* has been tabulated in Table 3. From the table it can be seen that above 80 percent of the roots were in the top foot of soil, which is the most important in controlling erosion.

Some species shed many leaves, which collect and conserve moisture in addition to preventing erosion. A layer of dry leaves up to 4 inches deep may be deposited in a single season; these rot down and fill small ditches and thus prevent deterioration of the land.

In controlling landslides and keeping flooded rivers within their courses, as well as in slowing down the speed of flow, bamboo may be used in either or both of two ways: as strong flexible fencing of cut bamboo, or as growing rows of plants with firm root systems.

Cultivation in the Garden. When bamboo is young, it should be left uncut. This will encourage strong bamboo shoots to sprout every year and will enhance the quality of the leaves.

TABLE 3. NUMBER AND DISTRIBUTION OF ROOTS OF BAMBUSA TULDA

Distance from Clump in Feet	Number of Roots at Specified Depth		
	0–1 ft.	1–2 ft.	2–4 ft.
0–1	1,300	33	18
1–2	1,100	22	14
2–3	1,200	44	12
3–4	1,020	61	25
4–6	875	143	49
6–8	385	207	62
8–10	228	142	69
10–14	388	145	103
Over 14	606	199	85
Totals	7,102	996	437

However, the older culms should be removed annually. They are easily damaged by insects, and they also inhibit new shoots from appearing. Thus the bamboo grove should be cut selectively every autumn: the large-culmed plants should be cut when they reach 5 or 6 years old and the smaller culms at 3 or 4 years.

Such general rules apply more particularly to the commercial management of a bamboo grove for income. In a garden, the cutting may be delayed by a year or more. When the older bamboos are allowed to remain, the number of new plants is reduced, but they tend to be larger in diameter.

Severe pruning of individual bamboo culms is, unlike the pruning of other plants, discouraging to their growth and beauty. Moso may be pruned at its top in moderation; this should be carried out when growth of the shoot is nearly complete and when one or two branches are beginning to appear from the lower part of the culm. Once this is done, a profusion of fresh leaves will grow in the following year.

Moso and madake are esteemed when they are large in diameter. To produce culms of large diameter, the largest culms should be left uncut since it is their characteristics that are transmitted to the next generation. In selecting culms to produce those of the largest growth, it is common sense also to excise the small or the unhealthy bamboo. Fertilization is essential to promote the healthiest growth and an increase in diameter.

Bamboo is an active and spreading plant. Unless it is inhibited, it will extend its growth over a large area. To keep it under control in the garden and growing as the cultivator wishes, the following measures will generally be found effective:

- Stop the spreading of roots by burying concrete slabs in the way of the rhizomes. The slabs should be 1 to 2 inches thick and 3 feet wide.
- Cut off immediately shoots which appear in undesired places.
- Frame the bamboo grove with pebbled pathways. By thus hardening the soil, fewer shoots will reach the surface and rhizome growth will be discouraged.
- The easiest way of keeping bamboo within bounds is to select the sympodial species which stay in clumps.

The elegance of the decorative thin-culmed species of bamboo is enhanced when they are short in height, and it is often necessary to check their growth. With the slender black bamboo (*Phyllostachys nigra*), for example, the methods, are as follows:

- Always leave the 1-year-old culms. Cut down the 2-year-old plants in the autumn when the new shoots have completed their growth.
- Do not prune the culms hard at the top; this spoils their naturalness and beauty. However, to make the parent bamboo produce next season numerous culms which are small, the cultivator may pull off some of its leaves and in this way weaken it. If the leaves become discolored, a small amount of bone dust or oil cake in water solution should be applied to the roots.
- The culms of black bamboo remain green until late summer, when growth is normally completed. They become darker in the autumn. To achieve the best black color they should be grown in sandy soil with good drainage and also have considerable exposure to the sun.

To reach the desired balance between bamboo and other plants and trees it is important to choose the right species for the garden. It is often well to borrow the Eastern method of growing bamboo among rocks and moss. For example, the slender black bamboo standing surrounded by white sand and the delicate green of hair moss presents a spectacle of immediate appeal.

Fertilization. The quality and yield is much improved by fertilization; the culms become bigger and the healthiest dark green leaves appear. Bamboo consumes a considerable amount of inorganic nutrients, which are supplied in the normal course of events through the soil or by rain. But under intensive cultivation, not enough are naturally available and artificial supplements are necessary.

Nitrogen is the element most needed by bamboo, followed by potassium, phosphate, and silicate. Experiments show that the most effective proportions for application are 9 to 10 parts of nitrogen, 5 parts of phosphate, 5 parts of potassium, and 6 parts of silicate. The ideal quantity is about 80 pounds of nitrogen per acre, with the other elements in proportion as given above. Nitrogen is of course available in many kinds of fertilizer, both synthetic and natural, including farmyard manure and night soil. The harvest—i.e., the weight of the new culms—may be doubled by the correct use of fertilizers.

Silicate is necessary for growth and, depending on the species, forms at least 6 percent of the leaf weight. Bamboo leaves are therefore left on the ground after cutting.

When a concentrated fertilizer is used, it should be applied about a month before the sprouts appear above ground, and similarly before the best growing season of the rhizome. Thus for madake the best times begin in April and July.

Figure 4 shows the results of effective fertilizing, based upon controlled 2-year experiments in Kyoto.

FLOWERING

Most bamboo flowers once only in its life. After this, it usually dies.

The reasons for the death of bamboo are that the old leaves fall at the time of flowering and, instead of being swiftly followed by new ones, are generally replaced by flowers. Thus few green leaves remain and the ability of the bamboo to take in water and nourishment is drastically reduced. This restriction on its growth causes the plant to lose vigor and finally brings about its death. Some species with an abundance of leaves produce green leaves of a secondary type and do not die. There is no cause for the belief that bamboo's dying after it comes to flower derives from infection.

Since bamboo usually reproduces itself asexually—i.e., without bearing flowers and pollen which are fertilized by contact with other plants—its flowering stage marks the one brief period

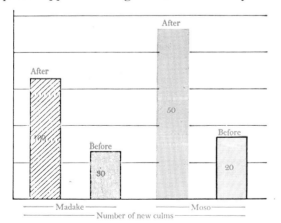 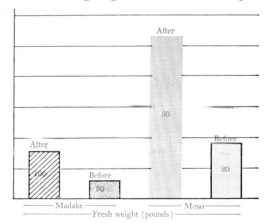

Fig. 4. Effects of Fertilizing

in its life when it can extend its growth sexually since some seeds are produced at this time.

About the phenomenon of flowering, there still remains considerable knowledge to be gathered. All the facts are by no means yet clear. There do exist some species that flower annually or almost annually, but these are not widespread. It is certain that the act of flowering occurs in general at very long intervals and also irregularly. The immediate cause of flowering is still a mystery. The most interesting symptom is that it usually occurs gregariously. That is, practically every bamboo of the same species, young or old and however widely separated they may be—even in different countries—will flower in or about the same year. This demonstrates, according to botanists, the common origin of these bamboos and the strength of the inherited habits of the genus.

The periods of flowering vary immensely. The most common Japanese bamboo, madake, flowers probably about every 120 years. Sasa species flower roughly every 60 years. Flowering cycles based upon observation of Indian sympodial bamboo are shown in Table 4, though the figures are probably too small.

TABLE 4. FLOWERING CYCLES

Schizostachyum species	30–40 years
Bambusa arundinacea	30–45 years
Bambusa polymorpha	55–60 years

When it flowers, the sympodial or clump-forming bamboo does so in the winter and generally produces seeds in the spring; the monopodial or single-culm species flower in the early summer and bear seeds in the autumn.

Generally, the clump-forming species produce fertile seed, from which they may be regenerated. Most of the single-culm species, however, do not produce fertile seeds, and the new grove has to come from rhizomes. This is an uncertain process, since, while the monopodial culm is almost sure to die within 2 years after flowering, the rhizome usually does also. Nevertheless, what will often occur is that—in the same year as the main bamboo flowers—very slender new culms grow from the original rhizome (see Figure 5). These slender culms flower also. The

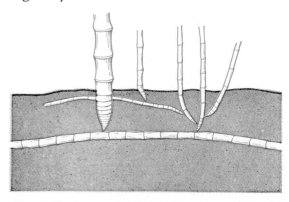

FIG. 5. REGENERATION AFTER FLOWERING. The old culm, which has flowered, is at the left; the slender new culms are at the right.

next year another small shoot appears (which may flower as well), but this grows from the new rhizome at the base of the slender culm of the previous season. In the subsequent year there is no flowering, and a new culm will sprout from the new rhizome system. By this time the new system should be independent, although it is still small and weak, so that when the parent rhizome dies, the bamboo has a good chance of survival.

If the bamboo grove is left without treatment after flowering, it will take over 10 years for the next family of culms to reach the size of the previous generation. But these steps can be taken to hasten revival:

- Those culms which have flowered (or are damaged by insects and disease) should be cut. If any have flowered and are healthy, they may be retained.
- The small new bamboo which emerges after flowering represents the foundation

of the new plant. It must be encouraged by every means and should be left uncut for the following 3 or 4 years.
- The new shoots should be adequately fertilized. Weeds in the bamboo grove must be kept under control.
- In order to avoid a further simultaneous flowering at a later date, shoots of a different stock of bamboo which has not flowered should introduced.

While it is quite possible to regrow sympodial or clump-forming bamboo from seeds—and this will happen naturally if the plants are left alone—it saves time in regenerating the grove if transplanted cuttings are brought to it.

INDUSTRIAL CULTIVATION

The world's yearly growth of bamboo is greater than 10,000,000 tons, of which almost all comes from the East. (It is estimated that in China alone there is a yearly growth of 3,500,000 tons.) However, far from all of this is actually used. Approximate figures of consumption are given in Table 5.

Its uses are legion but may be summarized into three categories: conversion to pulp for papermaking, utilization for various manufactured objects, and for food. Bamboo, having a high cellulose content, is well suited for making paper and also rayon—but only where cheap timber is not immediately available and where inexpensive labor may be secured. This is because the processing costs involved in producing and refining cellulose from bamboo are not small; and, where as basic a material as paper is being considered, the cost of raw materials is very significant. However, paper pulp is the main use to which the world turns bamboo. In India 800,000 tons of bamboo a year are converted into paper. East Pakistan processes nearly 90,000 tons; Thailand, Taiwan, Indonesia, and Burma account for another 100,000. These figures are expected to increase fairly soon with more intensive and more efficient cultivation. The figure for Japan is small since the higher cost of local processing and the relative abundance of native wood do not render bamboo pulp attractive from the commercial standpoint. But bamboo may well be mixed with wood in paper making since inferior culms can be economically used.

In Japan, over 80,000 tons of bamboo sprouts a year are used as food. (About the same weight is eaten in Taiwan, with one-sixth of the population.) The best culms are then employed industrially, and it is estimated that as many as 30,000 tons are exported in various fabricated forms.

The growing of bamboo is usually inefficient.

TABLE 5. YEARLY CONSUMPTION OF BAMBOO IN TONS

	TOTAL	PULP-MAKING	MANU-FACTURED GOODS	HOUSING	OTHERS
India	1,800,000	800,000	100,000	500,000	400,000
China	1,000,000	200,000	300,000	200,000	300,000
Burma	800,000	20,000	100,000	300,000	380,000
Japan	500,000	2,000	400,000	3,000	95,000
Indonesia	500,000	10,000	20,000	300,000	170,000
East Pakistan	490,000	90,000	20,000	300,000	80,000
Taiwan	350,000	60,000	250,000	20,000	20,000
Thailand	300,000	10,000	100,000	100,000	90,000

In almost every Eastern country the yield could be increased enormously by applying the simple rules of bamboo cultivation. But this is hardly ever done, mainly because bamboo is so easy to grow without attention that it is found everywhere in the East, and its real potential is overlooked. It occurs in myriad small holdings, difficult of access, uneconomic in size, and cultivated under marginal conditions by a farmer who is not receptive to change.

All the same, bamboo can be an attractive financial investment. Its average yearly increase in weight may be 20 percent. The period of growth is short and the cutting age is young. Further, bamboo demands little labor since there is no yearly planting or weeding strictly necessary. A study made in Japan to show the comparison in financial yields between a bamboo grove and a similar woodland cultivation is illustrated in Table 6.

At first glance this table seems to indicate that timber is a much more profitable crop than bamboo. Note, however, that the table does not take monetary interest into account. In timber cultivation, money is locked up for long periods of time, whereas bamboo produces profits yearly and hence becomes more commercially attractive, particularly for the small farmer with little capital.

In Thailand, as another example, bamboo is now used for construction and in manufacture, but its greatest future lies in the production of pulp for paper. The country's consumption of paper has tripled over the last 10 years to 150,000 tons, and less than one-quarter of this amount is made locally. Of this, only about 15 percent derives from bamboo at present. But the whole country's needs—and more—could be met from the bamboo forests in the north, given good cultivation and management.

In his *Bamboo in the U.S.A.*, W. H. Hodge estimates that bamboo annually produces 6 times as much cellulosic material per acre as does southern pine. If so, there might well be a good future for bamboo in papermaking in the West.

To cultivate a bamboo grove commercially is not difficult. If the accepted methods of bamboo husbandry are practiced, the yield can be multiplied considerably. These are summarized in the following paragraphs.

1) Suitable climate. Bamboo prefers a warm

TABLE 6. APPROXIMATE PROFITS PER ACRE FROM BAMBOO COMPARED TO TIMBER

	Figures given in U.S. dollars			
	BAMBOO (yearly crop rotation)		CRYPTOMERIA (40-year crop rotation)	
INCOME	Culms Culm sheaths	180 40 220	Thinnings Final cutting	1,800 5,500 7,300
EXPENDITURE	Fertilizer Tilling Harvesting, etc.	30 50 25 105	Land preparation Seedlings Thinning, weeding, etc.	50 80 170 300
PROFITS	Yearly return $115 Total after 40 years $4,600		Yearly return — Total after 40 years $7,000	

climate. For areas which have a severe winter, hardy species should be selected such as *Ph. henonis*, *Ph. pubescens*, *Semiarundinaria Kagamiana*.

2) Suitable soil and setting. Bamboo grows best in fertile soil which is well drained and mixed with gravel. But the best edible bamboo (moso) favors soil with a higher humidity and a greater proportion of clay. For optimum growth, water should be available in moderate amounts, neither too little nor to excess. Bamboo grows well on steep slopes.

Bamboo dislikes very strong sun. Sites facing the west and receiving the harsh afternoon heat are not ideal. In mild or warm climates, it is better to face the north; in cold regions, the south.

3) Fertilization. Fertilizer should be applied yearly to produce the best results. If artificial fertilizers are used, about 1,000 pounds per acre will be necessary; if natural manure is spread, the amount for good land should be about 7 times more. For poor land, these figures need to be increased.

4) Thinning. The most important item in managing a bamboo grove is spacing. Culms of good quality must have enough room for growth. An experiment carried out over 8 years with madake grown at different densities in an area of a quarter acre reinforces this point and is illustrated in Table 7.

TABLE 7. EFFECTS OF THINNING

Culms left after yearly cutting	1,200	900	750
Number of new culms yearly	165	179	210
Average weight in pounds	1,650	1,920	2,180

It will be seen that the number of new culms is higher in the lower-density grove, as is the weight of the year's crop. Since bamboo reproduces itself yearly, the optimum harvest or felling cycle is 1 year. Thus, depending on the rate at which the species reaches maturity, the 3- or 4-year-old culms are cut, giving a turnover of either 33 percent or 25 percent each year. However, a cycle which is yearly presupposes an abundance of labor and, as this is not always the case, longer cycles are often employed. With this wider spacing, the yield is relatively lower since the culms, having reached maturity, do not develop further but inhibit growth of the new sprouts and, in the case of an extremely long cycle, die.

5) Cutting and felling. Culms are cut with a hatchet or saw as close as possible to the ground. Diseased or otherwise useless plants should be cut even if they are young. The best method is to cut the good culms selectively according to their age. However, the simplest method of harvesting a bamboo grove is to cut down *all* the culms at one time, or clear-fell the area. This requires no supervision, but is a crude and drastic procedure. When it is done the supply of nutrition to the rhizome is suddenly decreased; this virtually arrests its growth, the number of good new shoots falls, and those that do appear are slender. These severe results can be mitigated by timing such clear-felling to allow the new shoots the best chance of growth. But without care it may take 10 years before the size of the newly grown bamboo reaches that of the original grove.

An improvement on this method is to clear-fell in belts of about 20 yards wide, since it seems that mature plants transmit nutrients to the new shoots through the rhizome for about half this length. Thus, shoots on either side of this belt have a chance of obtaining nutrition from the rhizomes beneath it. With fertilization, the newly grown bamboos regain the size of the parent in about 5 years.

Clump bamboo should be cut about 8 inches above the ground on a thinning principle so that the young and most healthy shoots which are left have support and the clump is kept open and workable. Cutting merely the periphery of the clump should be avoided.

6) Seasoning. The best age for cutting bamboo is about four years for large culms, and two years for thinner culms. The most suitable time is from October to December since bamboos cut at this time are less likely to be attacked by insects. When cutting off the branches, it is important that the hard skin of the culm shall not be peeled or scarred.

For fabrication, the cut bamboo should first be dried in the open for at least 20 days, preferably standing upright. If it is dried horizontally on the ground, double this time must be allowed.

The most important point in seasoning bamboo is to remove the oil from the culm. This serves two purposes. First, it cleans the skin and prevents it from becoming infected by mold or attacked by small insects; second, it preserves the culm by further hardening its exterior. The oil is traditionally removed by heating or steaming.

Heating is a simple process: a fire is made (or an electric heater employed) and the bamboo is placed in heat of about 200° F for ten minutes. Care must of course be taken not to burn the culm. As heat is applied, the oil oozes out and this is wiped off with a coarse rag. Steaming is carried out with boiling water or a solution of caustic soda and water which is boiled and its vapor run through long pipes; into these the bamboo is inserted for five or ten minutes and the exuded oil again wiped off quickly. After the oil has been removed by either process, the bamboo is washed in water and the alkalinity removed or reduced by rubbing with straw ash. The culms are straightened afterwards where necessary by being allowed to dry under tension in a frame.

Today new processes are also available to prevent mold and attack by insects. They are more effective and quicker, though more costly. Fumigation with diluted methyl bromide acts powerfully.

Selected culms may also be given a uniform appearance by scouring them with wet sand and exposing them to strong sunlight for about a day on each side.

7) Preparation for paper-pulp manufacture. Before processing, it is necessary to crush the bamboo and its nodes in rollers to aid chemical penetration and to free the large amount of air it holds.

Since bamboo contains about 20 percent starch, which normally stains pulp, this must be

TABLE 8. YEARLY YIELDS PER ACRE

Species	Tons per Acre (air dry weight)	
Phyllostachys pubescens *Ph. bambusoides* *Ph. nigra*	4.5 2.8 2.0	Japan (one-year cycle)
Dendrocalamus strictus *Melocanna baccifera* *Bambusa arundinacea*	1.2 2.0 1.6	India (three-year cycle)
B. stenostachya	1.2	Taiwan (seven-year cycle)
Thyrsostachys siamensis	1.0	Thailand (three-year cycle)

"digested" in two processes. In the first, a weak alkaline solution removes the starch before digesting with caustic soda and sodium sulphide for one hour at 300° F and two hours at 275° F. After this, the pulp is washed and bleached in the normal manner by use of sulphates.

The fresh-weight yield of bamboo varies considerably from one species to another; some typical yields are shown in Table 8. On an average, it takes 2.5 tons of fresh bamboo to produce 1 ton of dry pulp, but here too there is considerable variation according to the species used.

SPECIAL TECHNIQUES

Bonsai. The Japanese art of growing real trees in miniature in small pots can also be practiced with bamboo. But it is more difficult in view of the rapid rate of bamboo's growth, compared, say, with that of an oak.

The easiest bamboo to grow in pots is naturally the small bamboo grass or sasa. Of this species, the following are the prettiest: oroshimachiku (*Pleioblastus pygmaeus*), shakotanchiku (*Sasa palmata nebulosa*), kumazasa (*Sasa veitchii*), yadake (*Pseudosasa japonica*), and kanchiku (*Chimonobambusa marmorea*).

The smallest of these may be used by simply digging them up and transplanting them into pots. The larger types do not fit easily into small pots and should first be grown in a nursery to allow the new shoot to appear which will be trained for bonsai.

This is done in two stages. First, a 1-year-old culm and its shoots are grown in a nursery, having been previously transplanted in early spring. When the new shoots have completed their growth, the tall culms from the previous year should be cut. At this time, the smallest and most elegant of the shoots from the newly grown crop will be chosen, dug up carefully with their rhizomes and root hair, and planted in pots. These shoots will be about 12 inches high. It is customary to enhance the beauty of bonsai by the addition of rocks and moss. The best soil for use in bonsai is leaf mold, but a good substitute is red clay and fine sand mixed in equal proportions.

After sasa is planted in the pot, it should be watered about twice a day. If the leaves become discolored, a small amount of oil-cake fertilizer should be applied to the soil in a solution of water. If the culms begin to grow too much they

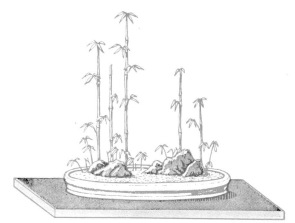

Fig. 6. A Bonsai of Bamboo

should be trimmed at the top by hand. But this is best done by breaking gently, not by cutting. The same discipline applies to the bonsai bamboo as to the normal one: the 2-year-old culms should be cut every autumn to make room for new.

The conditions for growing the larger types of bamboo in pots are very similar to those mentioned above, but they are more critical and even more care is needed. The main species of small bamboo suitable are: kurochiku (*Ph. nigra*), hoteichiku (*Ph. aurea*), narihiradake (*Semiarundinaria fastuosa*), and horaichiku (*Bambusa multiplex*).

A rhizome is selected and dug up in spring.

It should be young and have about 20 nodes with plenty of root hair, which must be disturbed as little as possible in cutting with a saw. One year in a nursery will usually be adequate to propagate a shoot and culm of a sufficiently small size. Sometimes, however, a single year is not enough, and it should be left for a second year to produce a plant of the right degree of smallness. To reduce the growth of a bamboo culm, it should be deprived of water for about a week during its period of fastest development.

When the shoots in the nursery have grown to about 4 inches in height, their sheaths should be peeled off. This is done by inserting scissors between the sheath and the culm—starting from the lowest node—and gently cutting the sheaths from the top down to the node into thin strips, which are then removed carefully and without damaging the culm. The sheaths and the strips must not be torn off by hand as this has a serious effect on the culm. After removal of the first sheath, several hours should be allowed before starting on the second, and so on. The treatment should also be followed through the evening, since bamboo continues its growth by night. Once the stripping operation has been completed, the bamboo bonsai should be given a little oil cake, or other compound fertilizer, in spring and in early summer to improve its color. The tender shoots without sheaths should initially be protected from the sun and must be watered if they get dry. Finally, they should be transplanted into the bonsai pot in the autumn after the new shoots have completed their growth. The rhizomes are cut to fit into the pot and a mixture of large and small stems planted to simulate the variety of a bamboo grove.

There is another method of controlling the growth of shoots, which is through repeated transplanting. First, the bamboo sprout and rhizome are dug up as described and planted in a nursery. When the shoots grow to a height of about 5 inches, they are then replanted. As soon as the shoot begins to pick up strength to recommence growth—in 7 to 10 days—it is transplanted another time. This may be carried out up to 3 times in a season. If it is desired to control growth even more, then the soil attached to the roots should be washed off at the time of transplanting, and/or the plants should be starved of water after repotting. They may also be put in the early sun for 2 to 3 hours, which is a treatment the bamboo does not greatly enjoy. The afternoon sun, however, may be too strong.

After the bamboo has reached the right height, it should be cared for in the normal way—watered, fertilized when necessary, and pruned of unneeded and old culms every year.

Square Bamboo. Bamboo may also be grown artificially in square section. Lengths in this shape are much sought after and used particularly in Japanese homes as an ornamental post for the tokonoma —the most important part of the living room.

Moso is usually selected because of the thickness of its culm. Square bamboo is produced by placing a frame around the sprout and forcing it to grow within the confines.

The frame is made from cedar wood about ½ inch thick and 10 to 12 feet long. The boards are cut into various widths, with one end somewhat narrower than the other to accommodate the natural shape of the bamboo. Two boards are nailed together at right angles to make half the frame and the 2 half-frames are tied together with a straw rope. The bottom of the frame is initially hung about 3 inches from the ground. When the apex of the bamboo shoot begins to show through the top, the frame should be slid up to produce a longer square length. The frame is removed when the bamboo reappears again and its sheaths begin to fall off, since this indicates that growth is complete and the square form will not change.

The size of the frame and the bamboo culm

must agree, since, if the bamboo is smaller than the walls of the frame, it will not become square. On the other hand if the bamboo is much larger than the frame, its circumference will be compressed too much and it will grow with vertical folds, not yielding a good square shape. It is therefore necessary when the bamboo shoot is about 1 foot above the ground to make an estimate as to how large it will grow in diameter before selecting an appropriate frame. The lower end of the frame should be a little larger than the top to allow the shoot to have some natural spread.

The production of rectangular bamboo is also

Fig. 7. Making Square Bamboo

perfectly possible through the use of a rectangular frame.

To attain square bamboo in a large size and of good quality, it is essential that a number of sturdy parent bamboos should be left uncut and not put into frames. If all the new bamboo shoots are made into square culms, there will be no parent bamboos left, and the new shoots from the rhizome in the following year will be slender and inferior.

In a bamboo grove of ¼ acre, about 200 parent bamboos should be left. Of the spring shoots, 100 or so should be selected for making into square bamboo and 50 of the best should be left as parent culms for subsequent years. The rest of the new sprouts may be dug up for eating or otherwise disposed of.

A pleasing and irregular pattern of brown spots on the yellow background may be permanently applied to square—or any—bamboo through artificial means. A solution known as chemical mud is painted on the culm with a brush. This mixture is left on the culm for the 3 or 4 months it remains in the ground after growth has finished and the frame has been taken away.

Chemical mud is made from 2 parts of a clay and water slurry plus 3 parts of 60-percent sul-

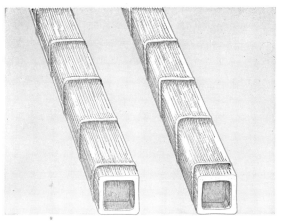

Fig. 8. Two Lengths of Square Bamboo

phuric or hydrochloric acid. A design of definite form may be produced if the white waxy powder which occurs on the culm at this stage is carefully removed and the mud then painted on. A stencil may be used if desired.

Natural markings also develop on bamboo and are usually the result of a fungus attack, but they confer an additional value as such culms are widely appreciated for decorative purposes.

These patterns can be engendered artificially. The variety known in the East as sesame-seed bamboo is produced by cutting off the upper parts of the culm and sawing halfway through the base in spring. This so weakens the remaining bamboo that fungus breeds vigorously on it until the culm is cut, decorating the entire surface with small black spots.

Square bamboo—with or without design—should be cut down between October and December of the year in which it has been formed. After it is cut and the branches have been lopped, the chemical solution should be washed off before the drying process. The culms are washed, using chaff and brushes; the nodes are given a scrubbing with a coarse brush made out of the small branches of bamboo. After the culms are dried, the oil should be removed and any curves straightened. With square bamboo, it is even more important to remove all oil in order to preserve a good color.

TABLE 9. SYMPODIAL SPECIES SUITABLE FOR TROPICAL AND SUBTROPICAL ZONES

Scientific Name	Alternate Scientific Name	Local Name	Height (Feet)	Diameter (Inches)	Description
Bambusa arundinacea Willd	*Bambos arundinacea* Retz.	*bans* or *kato* (India)	40–75	2.0–7.0	Generally known as thorny bamboo from the thorny branches of its thick culm. Grows well in wet soil, planted as windbreak around houses.
B. ventricosa McClure	*Leleba ventricosa* Lin	Buddha bamboo (Thailand)	5–20	0.5–2.0	Culm internodes are short and swollen; a most attractive ornamental plant.
Dendrocalamus giganteus Munro	*Bambusa gigantea* Wall.	*worra* (India)	50–90	6.0–12.0	The largest bamboo in the world with thick culms and numerous branches.
D. strictus Nees	*Bambos strictus* Roxb.	*bans kaban* (India)	15–40	1.0–3.0	Thick culm walls and curved upper branches. Grows well in dry soil.
Leleba multiplex Nakai	*Bambusa nana* Roxb.	*horaichiku* (Taiwan, Japan)	5–15	0.5–1.5	Thin-walled culms, numerous branches. Known as "hedge bamboo" in U.S.
L. multiplex forma Alphonso Karri Nakai	*B. multiplex forma Alphonso Karri* Nakai	*suhochiku* (Taiwan, Japan)	3–10	0.3–1.0	Elegant leaves and yellow culm with deep green stripes.
L. vulgaris var. striata Nakai	*B. striata* Loddiges	*pai sang kom* (Thailand)	15–40	1.5–4.0	Attractive ornamental plant with orange and yellow striped culm.
Melocanna baccifera Kurz	*Melocanna bambusoides* Trin.	*muli* (East Pakistan, India)	30–50	2.0–4.0	Culm walls thin, bears figlike fruit. Used for paper making.
Oxytenanthera nigrociliata Munro	*Oxytenanthera auriculata* Kurz	*kali* (East Pakistan)	30–50	2.0–4.0	Thin culm wall and slender branches. Used for papermaking.
Schizostachyum brachycladum Kurz		*buloh padi* (Malaysia)	25–40	1.5–4.0	Green or yellow culm with narrow stripes; an attractive ornamental plant.
S. zollingeri Steud.	*Schizostachyum chilianthum* Kurz	*buloh nipis* (Malaysia)	15–40	1.0–4.0	Thin culm wall. Used for papermaking.
Thyrsostachys oliveri Gamble		*pai ruak* (Thailand)	25–40	1.5–3.0	A graceful bamboo planted in the garden or along the road.

TABLE 10. MONOPODIAL AND INTERMEDIATE SPECIES SUITABLE FOR TEMPERATE ZONES

Scientific Name	Alternate Scientific Name	Japanese Name	Height (Feet)	Diameter (Inches)	Description
Arundinaria fastuosa Makino	*Semiarundinaria fastuosa* Makino	narihiradake	15–30	1.0–1.5	Tall, straight, and hardy. Culm a reddish purple, 3 branches from each node.
A. hindsii Munro	*Pleioblastus hindsii* Nakai	kanzanchiku	15–30	0.5–2.0	Many branches at top of culm. Likes deep shade.
A. japonica Sieb. et Zucc.	*Pseudosasa japonica* Makino	yadake	7–10	0.2–0.5	The hardiest *Arundinaria*, common in Japan. The very straight branches, 1 per node, are used for arrows.
A. pygmaea Makino	*Pleioblastus pygmaeus* Nakai	oroshimachiku	1.0–1.5	0.1	The smallest bamboo in the world.
Phyllostachys bambusoides Sieb. et Zucc.	*Phyllostachys reticulata* K. Koch	madake	30–60	1.0–5.0	Hardy with large leaves, 2 branches per node. Most widely spread in Japan, also common in China.
Ph. castillonis Mitford	*Ph. bambusoides var. castillonis* Makino	kimmeichiku	15–30	1.0–3.0	Hardy. Culms and branches bright gold with green stripes.
Ph. henonis Bean	*Ph. puberula* Makino	hachiku	30–45	1.0–4.0	Delicate in appearance but very hardy. Culm initially bright green. Lower nodes have white waxy powder.
Ph. mitis Bean	*Ph. pubescens* Mazel	moso	30–60	3.0–7.0	Node not prominent with single ring only. Sprout best for food. Largest bamboo in Japan, most widely spread in China.
Ph. mitis var. heterocycla Makino	*Ph. heterocycla* Matsum.	kikkochiku or butsumenchiku	15–30	2.0–4.0	Lower part of culm has alternating nodes shaped like tortoise shell.
Ph. nigra Munro	*Ph. nigripes* Hayata	kurochiku	7–25	0.5–1.0	Culms become good black color when mature. Initially nodes have marked bloom of white powder.
Sasa veitchii Rehd.	*Arundinaria veitchii* Brown	kumazasa	2–5	0.1–0.3	Small, hardy species with 1 long branch per node. Thick leaves whiten at edge in autumn.
Shibataea kumasaca Makino	*Shibataea ruscifolia* Makino	okamezasa or bungozasa	3–6	0.1–0.2	Culms almost solid, of elliptical shape with 5 branches from each node.
Tetragonocalamus quadrangularis Nakai	*Chimonobambusa quadrangularis* Makino	shihochiku or shikakuchiku	10–15	0.5–1.0	Culms have 4 corners, squarish section. Nodes wide and surrounded by buds.

Photo Notes

The photographs in this book were made with a number of cameras and a variety of films and lenses. Most of the large color and "studio" shots were taken with a 4×5 Graphic with a f5.6 Ektar lens. A 6×6 Mamiyaflex with a 105mm f5.6 lens and a Rolleiflex with an 80mm f2.8 lens were used for the bulk of the color photos, as their versatility and relatively large format made them practical for "location" shooting. All the photos of the craftsmen in the final section of the book were shot with a Nikon F8s and a Nikkomat equipped with 24, 35, 50, and 200mm lenses. A number of outdoor shots were also taken with the Nikon equipment.

All the indoor 4×5 photos used Ektachrome type B film. The 4×5 black and white film was Fuji SS and SSS. The 6×6 color film in daylight type was Ektachrome X, while the 35mm color film was predominantly Kodachrome II. The black and white 35mm photos were shot on Kodak Plus X, Tri-X, and Recording film. The great speed of the latter (ASA 800) made it ideal for the many shots that had to be made with poor light conditions.

About half the photos were taken in the Tokyo area and half in the Kansai (Kyoto-Nara). Some few were also shot in Hong Kong, Cambodia, and Bangkok. The number of photographs before editing to the present 300-odd was in the neighborhood of 5,000.

Bibliography

Camus, E.G. *Les Bambusées*. Paris: Lechavier, 1913

de Fonblanque, E.B. *Niphon and Pe-che-li or Two Years in Japan and Northern China*. London: Saunders, Otley, and Co., 1863

Forest Research Institute. *Annotated Bibliography on Bamboo*. Dehra Dun, India: Forest Research Institute Press, 1960

Freeman-Mitford, A.B. *The Bamboo Garden*. London: Macmillan, 1896

Gamble, J.S. *The Bambuseae of British India*. New York: Johnson Reprint Corporation, 1896

Lawson, A.H. *Bamboo: A Gardener's Guide*. London: Faber, 1968

McClure, F.A. *The Bamboos: A Fresh Perspective*. Cambridge, Mass.: Harvard University Press, 1966

Spörry, H. *Die Verwendung des Bambus in Japan*. Zurich: Zürcher & Furrer, 1903

Ueda, K. *Studies on the Physiology of Bamboo*. Kyoto: Kyoto University Press, 1960

United States Department of Agriculture. *Bamboo in the United States* (Handbook No. 193). Washington, D.C.: Government Printing Office, 1961

———. *Growing Ornamental Bamboo* (Handbook No. 76). Washington, D.C.: Government Printing Office, 1963

Worcester, G.R.G. *Sail and Sweep in China*. London: Her Majesty's Stationery Office, 1966

Acknowledgments

Many thanks are due to Mr. Ryo Sato, proprietor of the excellent Living Arts and Crafts shop in Yokohama, who, as an ardent admirer of bamboo himself, gave the initial encouragement for this book and also supplied much information.

A considerable debt of gratitude is owed to Miss Sachiko Suzuki for her great patience and tact in securing interviews and eliciting facts, to say nothing of the many hours she spent carrying heavy camera equipment under the hot sun.

For permission to photograph various objects from a number of collections, thanks are due to Walter Lutz, Meredith Weatherby, Mrs. John Jerwood, Michael Chang, and Eric Trigg; the Atami Art Museum; the National Palace Museum, Taiwan; the Bamboo Museum, Iidabashi, Tokyo; and the Tokyo antique shops of Honma and Yamada.

A word of explanation should be added concerning the division of responsibilities among the three authors. The original concept came from Robert Austin, a graphics consultant and book editor for the *Reader's Digest*, Tokyo. He obtained the cooperation of Koichiro Ueda, professor emeritus of Kyoto University, president of the Japan Bamboo Industries Association, and Japan's leading authority on bamboo; and Dana Levy, photographer, graphic designer, and a creative director of the Tokyo offices of McCann Erickson-Hakuhodo. The entire text is the work of Mr. Austin, most technical information having been supplied by Professor Ueda. All the photographs and the layout are Mr. Levy's. The typography and book design are the work of Mr. Austin, Mr. Levy, and the publisher.

In the fullest sense, then, this book represents a close collaboration between the three authors. Their joint hope and aim has been to portray bamboo, in its many guises, in an equal partnership of words and photographs.

The "weathermark" identifies this book as having been planned and produced at the Tokyo offices of John Weatherhill, Inc. Book design and typography by Robert Austin, Dana Levy, and Meredith Weatherby. Layout by Dana Levy. Composition by General Printing Company, Yokohama. Plates engraved and printed (in 5-color offset and monochrome gravure), together with the text, by Nissha Printing Company, Kyoto. Bound at the Okamoto Binderies, Tokyo. The type face used throughout is Monotype Bell.